hands-on science

Level Three

Jennifer Lawson

Joni Bowman

Randy Cielen

Carol Pattenden

Rita Platt

Winnipeg • Manitoba • Canada

Peguis Publishers acknowledges the financial support of the Government of Canada through the Book Publishing Industry Development Program (BPIDP) for our publishing activities.

Canadä

00 01 02 03 04 5 4 3 2 1

Canadian Cataloguing in Publication Data

Main entry under title:

Hands-on science : level three

Ont. ed.
Includes bibliographical references.
ISBN 1-894110-46-3

1. Science – Study and teaching (Primary).
   I. Lawson, Jennifer E. (Jennifer Elizabeth), 1959-

LB1532.L386 2000 372.3'5044 C00-920016-9

Series Editor: Leigh Hambly
Assistant Editor: Catherine Lennox
Book and Cover Design: Suzanne Gallant
Illustrations: Pamela Dixon, Jess Dixon

Program Reviewers

- Karen Boyd, Grade 3 teacher, Winnipeg, Manitoba
- Peggy Hill, mathematics consultant, Winnipeg, Manitoba
- Nancy Josephson, science and assessment consultant, Ashcroft, British Columbia
- Denise MacRae, Grade 2 teacher, Winnipeg, Manitoba
- Gail Ruta-Fontaine, Grade 2 teacher, Winnipeg, Manitoba
- Judy Swan, Grade 1 teacher, Winnipeg, Manitoba
- Barb Thomson, Grade 4 teacher, Winnipeg, Manitoba.

100-318 McDermot Avenue
Winnipeg, Manitoba, Canada R3A 0A2

Email: books@peguis.com
Tel: 204-987-3500 • Fax: 204-947-0080
Toll free: 1-800-667-9673

Contents

Introduction 1

Assessment 13

Unit 1: Growth and Changes in Plants 27

Books for Children 28
Web Sites 29
Introduction 30
1 Parts of a Plant 32
2 Classifying Plants 34
3 Special Features of Plants 37
4 Life Cycle of a Plant 44
5 Plants Need Water and Light 47
6 The Effects of Soil and Nutrients on Plants 54
7 Plants and Seasonal Changes 57
8 Plants That We Eat 64
9 Where Does Your Garden Grow? 65
10 Plants and Their Importance to Humans 68
11 Making Dye From Plants 70
12 Plants and Animals Working Together 73
13 Plants and Soil Erosion 76
14 Protecting Plants 79
References for Teachers 80

Unit 2: Magnetic and Charged Materials 81

Books for Children 82
Web Sites 83
Introduction 84
1 Magnetic Attraction 85
2 Magnets in Everyday Life 89
3 Making a Magnet 92
4 Magnetic Force 94
5 Designing and Constructing With Magnets 100
6 Games With Magnets 103
7 Magnetic Fields and Polarity 106
8 Static Electricity 108
9 Static Electricity and Humidity 111
10 The Force of Static Electricity 114
11 Making an Electroscope 117
12 Static Electricity in Everyday Life 121
References for Teachers 124

Unit 3: Forces and Movement 125

Books for Children 126
Web Sites 127
Introduction 128
1 Forces – Push or Pull 129
2 Friction 132
3 Magnetic Force 135
4 Static Electrical Force 138
5 Gravitational Force 140
6 Energy and Movement 144
7 Designing Toys That Use Different Forms of Energy to Move 147
8 Everyday Devices 151
9 Parts of a System 153
References for Teachers 156

Unit 4: Stability 157

Books for Children 158
Web Sites 159
Introduction 160
1 Altering the Strength of Objects 161
2 Levers 164
3 Balance Points 170
4 Types of Bridge Structures 173
5 Bridge Construction 176
6 Designing and Constructing Bridges 183
7 Structure and Mechanical Parts of Objects 186
References for Teachers 188

Unit 5: Soils in the Environment 189

Books for Children 190
Web Sites 191
Introduction 192
1 Different Types of Soil 194
2 Components of Soil 197
3 Absorption of Water 200
4 How Different Soils Affect the Growth of Plants 203
5 The Characteristics of Soils and Plants 206
6 Investigating Living Things Found in Soil 210
7 The Effect of Moving Water on Soil 213
8 The Effects of Rainfall on Surfaces 216
9 Recycling Organic Materials 219
10 Using Soil Materials to Make Useful Objects 221
References for Teachers 224

Introduction

Program Introduction

Hands-On Science develops students' scientific literacy through active inquiry, problem solving, and decision making. With each activity in the program, students are encouraged to explore, investigate, and ask questions as a means of heightening their own curiosity about the world around them. Students solve problems through firsthand experiences, and by observing and examining objects within their environment. In order for young students to develop scientific literacy, concrete experience is of utmost importance – in fact, it is essential.

The Foundations of Scientific Literacy

Hands-On Science focuses on the four foundation statements for scientific literacy, as established in the *Pan-Canadian Protocol.** These foundation statements are the bases for the learning outcomes identified in ***Hands-On Science.***

Foundation 1: Science, Technology, Society, and the Environment (STSE)

Students will develop an understanding of the nature of science and technology, of the relationships between science and technology, and of the social and environmental contexts of science and technology.

Foundation 2: Skills

Students will develop the skills required for scientific and technological inquiry, for solving problems, for communicating scientific ideas and results, for working collaboratively, and for making informed decisions.

Foundation 3: Knowledge

Students will construct knowledge and understandings of concepts in life science, physical science, and earth and space science, and apply these understandings to interpret, integrate, and extend their knowledge.

Foundation 4: Attitudes

Students will be encouraged to develop attitudes that support responsible acquisition and application of scientific and technological knowledge to the mutual benefit of self, society, and the environment.

**Common Framework of Science Learning Outcomes K-12: Pan-Canadian Protocol for Collaboration on School Curriculum (1997).*

Hands-On Science Expectations

UNIT 1: GROWTH AND CHANGES IN PLANTS

- ☐ Identify the major parts of plants and describe their basic functions.
- ☐ Classify plants according to visible characteristics.
- ☐ Describe, using their observations, the changes that plants undergo in a complete life cycle.
- ☐ Describe, using their observations, the effects of the seasons on plants.
- ☐ Compare the life cycles of different kinds of plants.
- ☐ Identify traits that remain constant in some plants as they grow.
- ☐ Describe, using their observations, how the growth of plants is affected by changes in environmental conditions.
- ☐ Explain how different features of plants help them survive.
- ☐ Design and conduct a hands-on inquiry into seed germination or plant growth.
- ☐ Ask questions about and identify some needs of plants, and explore possible answers to these questions and ways of meeting these needs.
- ☐ Plan investigations to answer some of these questions or find ways of meeting these needs, and explain the steps involved.
- ☐ Use appropriate vocabulary in describing their investigations, explorations, and observations.
- ☐ Record relevant observations, findings, and measurements, using written language, drawings, charts, and graphs.
- ☐ Communicate the procedures and results of investigations for specific purposes and to specific audiences, using drawings, demonstrations, simple media works, and oral and written descriptions.
- ☐ Describe ways in which humans use plants for food, shelter, and clothing.
- ☐ Describe ways in which humans can protect natural areas to maintain native plant species.
- ☐ Identify the parts of a plant that are used to produce specific products for humans.
- ☐ Describe various plants used in food preparation and identify places where they can be grown.
- ☐ Describe various settings in which plant crops are grown.
- ☐ Describe ways in which plants and animals depend on each other.
- ☐ Compare the requirements of some plants and animals, and identify the requirements that are common to all living things.
- ☐ Demonstrate awareness of ways of caring for plants properly.
- ☐ Identify some functions of different plants in their local area.

UNIT 2: MAGNETIC AND CHARGED MATERIALS

- ☐ Classify, using their observations, materials that are magnetic and not magnetic, and identify materials that can be magnetized.
- ☐ Identify, through observation, the effect of different conditions on the strength of magnets and on static electric charges in materials.
- ☐ Compare different materials by measuring their magnetic strength or the strength of their electric charge.
- ☐ Identify, through observation, pairs of materials that produce a charge when rubbed together.
- ☐ Describe and demonstrate how some materials that have been electrically charged or magnetized may either push or pull similar materials.
- ☐ Determine, through observation, the polarity of a magnet.
- ☐ Identify materials that can be placed between a magnet and an attracted object without diminishing the strength of the attraction.
- ☐ Predict, verify, and describe the interaction of two objects that are similarly charged.
- ☐ Describe, through observation, changes in the force of attraction at different distances, both for magnetic forces and for static electric forces.
- ☐ Design and construct a system that uses magnetic force to move an object.
- ☐ Ask questions about and identify problems related to magnetic and static electric forces, and explore possible answers or solutions.
- ☐ Plan investigations to answer some of these questions or solve some of these problems, and explain the steps involved.

▶

- ☐ Use appropriate vocabulary in describing their investigations, explorations, and observations.
- ☐ Record relevant observations, findings, and measurements, using written language, drawings, charts, and graphs.
- ☐ Communicate the procedures and results of investigations for specific purposes and to specific audiences, using demonstrations, drawings, simple media works, and oral and written descriptions.
- ☐ Identify uses of magnets in familiar things.
- ☐ Describe examples of static electricity encountered in everyday activities.
- ☐ Identify ways in which static electricity can be used safely or avoided.

UNIT 3: FORCES AND MOVEMENT

- ☐ Identify force as a push or pull by one body on another.
- ☐ Investigate the ways in which different forces can change the speed or direction of a moving object.
- ☐ Investigate the effect of magnets and electrically charged objects on the motion of different materials.
- ☐ Identify, through observation, different forms of energy and suggest how they might be used to provide power to devices and to create movement.
- ☐ Distinguish between kinds of motion and indicate whether the motion is caused indirectly or directly.
- ☐ Investigate the effects of directional forces and how unbalanced forces can cause visible motion in objects that are capable of movement.
- ☐ Ask questions about and identify needs and problems related to the behaviour of different forces in their immediate environment, and explore possible answers and solutions.
- ☐ Plan investigations to answer some of these questions or solve some of these problems, and explain the steps involved.
- ☐ Use appropriate vocabulary in describing their investigations, explorations, and observations.
- ☐ Record relevant observations, findings, and measurements, using written language, drawings, charts, and graphs.
- ☐ Communicate the procedures and results of investigations for specific purposes and to specific audiences, using drawings, demonstrations, simple media works, and oral and written descriptions.
- ☐ Design and construct a device that uses a specific form of energy in order to move.
- ☐ Describe the visible effects of forces acting on a variety of everyday objects.
- ☐ Identify surfaces that affect the movement of objects by increasing or reducing friction.
- ☐ Demonstrate how a magnet works and identify ways in which magnets are useful.
- ☐ Recognize devices that are controlled automatically, at a distance, or by hand.
- ☐ Identify parts of systems used in everyday life, and explain how the parts work together to perform a specific function.

UNIT 4: STABILITY

- ☐ Describe, using their observations, ways in which the strength of different materials can be altered.
- ☐ Describe ways in which forces alter the shape or strength of different structures.
- ☐ Describe ways to improve the strength and stability of a frame structure.
- ☐ Describe, using their observations, the role of struts and ties in structures under load.
- ☐ Describe, using their observations, the changes in the amount of effort needed to lift a specific load with a lever when the position of the fulcrum is changed.
- ☐ Describe, using their observations, how simple levers amplify or reduce movement.
- ☐ Describe the effects of different forces on specific structures and mechanisms.
- ☐ Ask questions about and identify needs and problems related to structures and mechanisms in their immediate environment, and explore possible answers and solutions.
- ☐ Plan investigations to answer some of these questions or solve some of these problems, and explain the steps involved.

▶

- ☐ Use appropriate vocabulary to describe their investigations, explorations, and observations.
- ☐ Record relevant observations, findings, and measurements, using written language, drawings, charts, and graphs.
- ☐ Communicate the procedures and results of investigations for specific purposes and to specific audiences, using demonstrations, drawings, simple media works, and oral and written descriptions.
- ☐ Design and make a stable structure that will support a given mass and perform a specific function.
- ☐ Use appropriate materials to strengthen and stabilize structures that they have designed and made and that are intended to support a load.
- ☐ Design and make a levered mechanism.
- ☐ Design and make a stable structure that contains a mechanism and performs a function that meets a specific need.
- ☐ Use appropriate equipment and adhesives when making structures that they have designed themselves.
- ☐ Use hand tools and equipment appropriately to cut a variety of materials.
- ☐ Distinguish between the structure of an object and its mechanical parts.
- ☐ Recognize that geometrical patterns in a structure contribute to the strength and stability of that structure.
- ☐ Demonstrate awareness that the strength in structures is due to bulk (or mass), number of layers, and shape.
- ☐ Identify a number of common levers and describe how they make work easier.
- ☐ Identify efficient ways of joining the components of a mechanical structure or system.
- ☐ Describe, using their observations, how different balance points of different masses affect the stability of a structure.
- ☐ Predict which body positions provide the most stability in various circumstances.

UNIT 5: SOILS IN THE ENVIRONMENT

- ☐ Describe, using their observations, the various components within a sample of soil.
- ☐ Describe, using their observations, the differences between sand, clay, humus, and other soil components, and compare and describe soil samples from different locations.
- ☐ Compare the absorption of water by different earth materials, and describe the effects of moisture on characteristics of the materials.
- ☐ Describe, using their observations, how different earth materials are affected by moving water.
- ☐ Compare different ways in which plant roots grow through the soil; describe through experimentation how soil can be separated into its different components.
- ☐ Ask questions about organisms and events in the outdoor environment and identify needs of organisms that arise from these events, and explore possible answers to these questions and ways of meeting these needs.
- ☐ Plan investigations to answer some of these questions or find ways of meeting these needs, and explain the steps involved.
- ☐ Use appropriate vocabulary in describing their investigations, explorations, and observations.
- ☐ Record relevant observations, findings, and measurements, using written language, charts, and drawings.
- ☐ Communicate the procedures and results of investigations for specific purposes and to specific audiences, using drawings, demonstrations, simple media works, and oral and written descriptions.
- ☐ Identify living things found in the soil.
- ☐ Demonstrate awareness of the importance of recycling organic materials in soils.
- ☐ Recognize the importance of understanding different types of soil and their characteristics.
- ☐ Describe how the use of different soils affects the growth of indoor plants.
- ☐ Describe ways of using soil materials to make useful objects, and investigate, through manipulation, ways of shaping clay to make useful objects.

Program Principles

1. Effective science programs involve hands-on inquiry, problem solving, and decision making.
2. The development of students' skills, attitudes, knowledge, and understanding of STSE issues form the foundation of the science program.
3. Children have a natural curiosity about science and the world around them. This curiosity must be maintained, fostered, and enhanced through active learning.
4. Science activities must be meaningful, worthwhile, and relate to real-life experiences.
5. The teacher's role in science education is to facilitate activities and encourage critical thinking and reflection. Children learn best by doing, rather than by just listening. The teacher, therefore, should focus on formulating and asking questions rather than simply telling.
6. Science should be taught in correlation with other school subjects. Themes and topics of study should integrate ideas and skills from several core areas whenever possible.
7. The science program should encompass, and draw on, a wide range of educational resources, including literature, nonfiction research material, audio-visual resources, technology, as well as people and places in the local community.
8. Assessment of student learning in science should be designed to focus on performance and understanding, and should be conducted through meaningful assessment techniques carried on throughout the unit of study.

Program Implementation

Program Resources

Hands-On Science is arranged in a format that makes it easy for teachers to plan and implement.

Units are the selected topics of study for the grade level. The units relate directly to the learning expectations, which complement those established in the *Pan-Canadian Protocol* and *The Ontario Curriculum, Grades 1-8: Science and Technology, 1998* documents. The units are organized into several activities. Each unit also includes books for children, a list of annotated web sites, and references for teachers.

The introduction to each unit summarizes the general goals for the unit. The introduction provides background information for teachers, and a complete list of materials that will be required for the unit. Materials include classroom materials, equipment, visuals, and reading materials.

Each unit is organized into topics, based on the expectations. The topics are arranged in the following format:

Science Background Information for Teachers: Some topics provide teachers with the basic scientific knowledge they will need to present the activities. This information is offered in a clear, concise format, and focuses specifically on the topic of study.

Materials: A complete list of materials required to conduct the main activity is given. The quantity of materials required will depend on how you conduct activities. If students are working individually, you will need enough materials for each student. If students are working in groups, the materials required will be significantly reduced. Many of the identified items are for the teacher to use for display ▶

purposes, or for making charts for recording students' ideas. In some cases, visual materials – large pictures, sample charts, and diagrams – have been included with the activity to assist the teacher in presenting ideas and questions, and to encourage discussion. You may wish to reproduce these visuals, mount them on sturdy paper, and laminate them so they can be used for years to come.

Activity: This section details a step-by-step procedure, including higher-level questioning techniques, and suggestions, for encouraging exploration and investigation.

Activity Sheet: The reproducible activity sheets are designed to correlate with the expectations of the activity. Often, the activity sheets are to be used *during* the activity to record results of investigations. At other times, the sheets are to be used as a *follow-up* to the activities. Students may work independently on the sheets, in small groups, or you may choose to read through the sheets together and complete them in a large-group setting. Activity sheets can also be made into overheads or large experience charts. Since it is important for students to learn to construct their own charts and recording formats, you may want to use the activity sheets as examples of ways to record and communicate ideas about an activity. Students can then create their own sheets rather than use the ones provided.

Note: Activity sheets are meant to be used only in conjunction with, or as a follow-up to, the hands-on activities. The activity sheets are not intended to be the science lesson itself or the sole assessment for the lesson.

Extension: Included are optional activities to extend, enrich, and reinforce the expectations.

Activity Centre: Included are independent student activities that focus on the expectations.

Assessment Suggestions: Often, suggestions are made for assessing student learning. These assessment strategies focus specifically on the expectations of a particular activity topic (assessment is dealt with in detail on page 13). Keep in mind that the suggestions made within the activities are merely ideas to consider – you may use your own assessment techniques, or refer to the other assessment strategies on pages 15-25.

Classroom Environment

The classroom setting is an important aspect of any learning process. An active environment, one that gently hums with the purposeful conversations and activities of students, indicates that meaningful learning is taking place. When studying a specific topic, you should display related objects and materials, student work, pictures and posters, graphs and charts made during activities, and summary charts of important concepts taught and learned. An active environment reinforces concepts and skills that have been stressed during science activities.

Time Lines

No two groups of students will cover topics and material at the same rate. Planning the duration of units is the responsibility of the teacher. In some cases, the activities will not be completed during one block of time and will have to be carried over. In other cases, students may be especially interested in one topic and may want to expand upon it. The individual needs of the class should be considered; there are no strict time lines involved in ***Hands-On Science***. It is important,

however, to spend time on every unit in the program so that students focus on all of the curriculum expectations established for their grade level.

Classroom Management

Although hands-on activities are emphasized throughout this program, the manner in which these experiences are handled is up to you. In some cases, you may have all students manipulating materials individually; in others, you may choose to use small-group settings. Small groups encourage the development of social skills, enable all students to be active in the learning process, and mean less cost in terms of materials and equipment.

Occasionally, especially when safety concerns are an issue, you may decide to demonstrate an activity, while still encouraging as much student interaction as possible. Again, classroom management is up to you, since it is the teacher who ultimately determines how the students in his or her care function best in the learning environment.

Science Skills: Guidelines for Teachers

While involved in the activities of ***Hands-On Science***, students will use a variety of skills as they answer questions, solve problems, and make decisions. These skills are not unique to science, but they are integral to students' acquisition of scientific literacy. The skills include initiating and planning, performing and recording, analyzing and interpreting, as well as communicating and the ability to work in teams. In the early years, basic skills should focus on science inquiry. Although the wide variety of skills are not all presented here, the following guidelines provide a framework to use to encourage students' skill development in specific areas.

Observing

Students learn to perceive characteristics and changes through the use of all five senses. Students are encouraged to use sight, smell, touch, hearing, and taste (when safe) to gain information about objects and events. Observations may be qualitative (by properties such as texture or colour), or quantitative (such as size or number), or both. Observing includes:

- gaining information through the senses
- identifying similarities and differences, and making comparisons
- sequencing events or objects

Exploring

Students need ample opportunities to manipulate materials and equipment in order to discover and learn new ideas and concepts. During exploration, students need to be encouraged to use all of their senses and observation skills. Oral discussion is also an integral component of exploration; it allows students to communicate their discoveries.

Classifying

This skill is used to group or sort objects and events. Classification is based on observable properties. For example, objects can be classified into living and nonliving groups, or into groups according to colour, shape, or size. One of the strategies used for sorting involves the use of Venn diagrams (either a double Venn or a triple Venn). Venn diagrams can involve distinct groups, or can intersect to show similar characteristics.

Venn Diagram With Distinctive Groups:

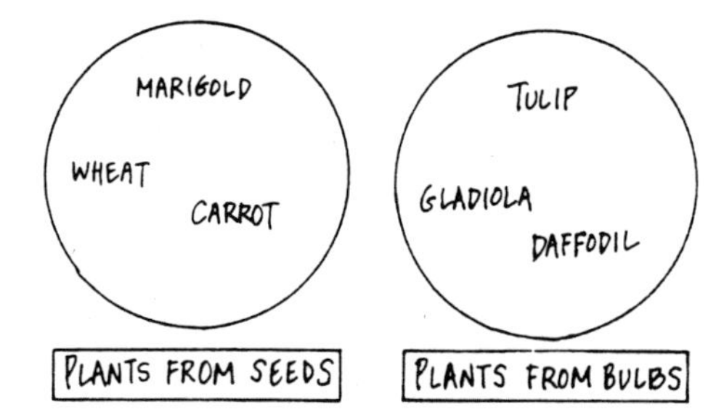

Intersecting Venn Diagram:

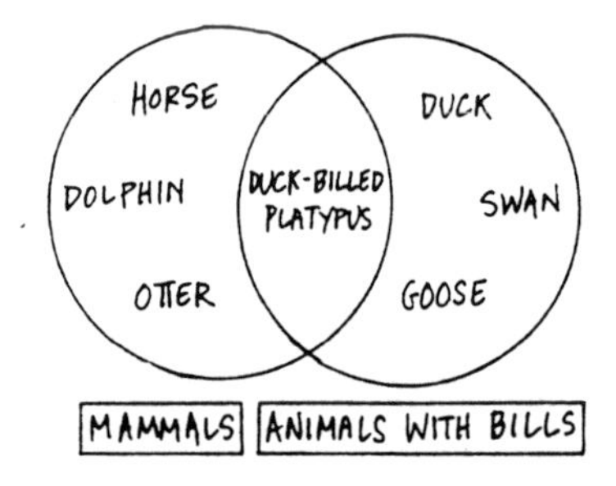

Measuring

This is a process of discovering the dimensions or quantity of objects or events and usually involves the use of standards of length, area, mass, volume, capacity, temperature, time, and speed. Measuring skills also include the ability to choose appropriate measuring devices, and using proper terms for direction and position.

In the early years, measuring activities first involve the use of *nonstandard* units of measure, such as unifix cubes or paper clips to determine length. This is a critical preface to measuring with standard units. Once standard units are introduced, the metric system is the foundation of measuring activities. Teachers should be familiar with, and regularly use, these basic measurement units.

An essential skill of measurement is *estimating*. Regularly, students should be encouraged to estimate before they measure, whether it be in nonstandard or standard units. Estimation allows students opportunities to take risks, use background knowledge, and learn from the process.

Length: Length is measured in metres, portions of a metre, or multiples of a metre. The most commonly used units are:

- millimetre (mm): about the thickness of a paper match
- centimetre (cm): about the width of your index fingernail
- metre (m): about the length of a man's stride
- kilometre (km): 1000 metres

Mass: Mass, or weight, is measured in grams, portions of a gram, or multiples of a gram. The most commonly used units are:

- gram (g): about the weight of a paper clip
- kilogram (kg): a cordless telephone weighs about 2 kilograms
- tonne (t): about the weight of a compact car

Note: When measuring to determine the heaviness of an object, the term *mass* is more scientifically accurate than the term *weight*. However, it is still acceptable to use the terms interchangeably in order for students to begin understanding the vocabulary of science.

Capacity: Capacity refers to the amount of fluid a container holds, and is measured in litres, portions of a litre, and multiples of a litre. The most commonly used units are:

- millilitre (ml): a soup spoon holds about 15 millilitres
- litre (l): milk comes in litre containers, or portions and multiples of a litre

Volume: Volume refers to the amount of space taken up by an object and is measured in cubic units, generally cubic centimetres (cm^3) and cubic metres (m^3).

Note: Volume and capacity are often used interchangeably. However, a teacher should use the terms correctly in context, referring to liquid measure as capacity and space taken up as volume. Early years students are not yet expected to master the differences in the concepts and terminology, and can, therefore, be allowed to use the terms *volume* and *capacity* interchangeably.

Area: Area is measured in square centimetres, or portions and multiples thereof. By becoming familiar with the units of length, the teacher can understand area measurements by thinking of that unit in a two-dimensional form, such as square centimetres (cm^2) and square metres (m^2).

Temperature: Temperature is measured in degrees Celsius (°C). 21°C is normal room temperature; water freezes at 0°C and boils at 100°C.

Communicating

In science, one communicates by means of diagrams, graphs, charts, maps, models, symbols, as well as with written and spoken languages. Communicating includes:

- reading and interpreting data from tables and charts
- making tables and charts
- reading and interpreting data from graphs
- making graphs
- making labelled diagrams
- making models
- using oral and written language

When presenting students with charts and graphs, or when students make their own as part of a specific activity, there are guidelines that should be followed.

- A *pictograph* has a title and information on one axis that denotes the items being compared. There is generally no graduated scale or heading for the axis representing numerical values.

Favourite Desserts

Cake **Pie** **Ice Cream**

- A *bar graph* is another common form of scientific communication. Bar graphs should always be titled so that the information communicated is easily understood. These titles should be capitalized in the same manner as one

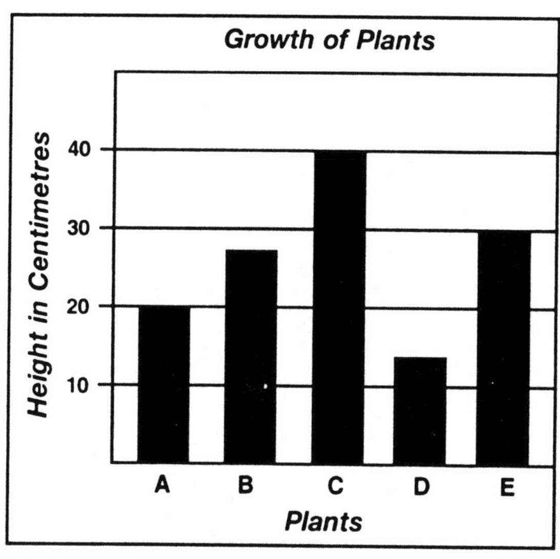

would title a story. Both axes of the graph should also be titled and capitalized in the same way. In most cases, graduated markings are noted on one axis and the objects or events being compared are noted on the other. On a bar graph, the bars must be separate, as each bar represents a distinct piece of data.

- *Charts* also require appropriate titles, and both columns and rows need specific headings. Again, all of these titles and headings require capitalization as in titles of a story. In some cases, pictures can be used to make the chart easier for young students to understand. Charts can be made in the form of checklists or can include room for additional written information and data.

Objects in Water		
Object	**Float**	**Sink**
[scissors drawing]		✓
[ball drawing]	✓	

Flowers We Saw		
Type	**Colour**	**Diagram**
daffodil	yellow	[drawing]
rose	red	[drawing]
lilac	purple	[drawing]

Measuring Length		
Object	**Estimate (cm)**	**Length (cm)**
book	30 cm	27 cm
pencil	10 cm	16 cm

Communicating also involves using the language and terminology of science. Students should be encouraged to use the appropriate vocabulary related to their investigations, for example, *objects, material, solid, liquid, gas, condensation, evaporation, magnetic, sound waves*, and *vibration*. The language of science also includes terms like *predict, infer, estimate, measure, experiment,* and *hypothesize*. Teachers should use this vocabulary regularly throughout all activities, and encourage their students to do the same. As students become proficient at reading and writing, they can also be encouraged to use the vocabulary and terminology in written form. Consider developing whole-class or individual glossaries whereby students can record the terms learned and define them in their own words.

Predicting

A prediction refers to the question: What do you think will happen? For example, when a balloon is blown up, ask students to predict what they think will happen when the balloon is placed in a basin of water. It is important to provide opportunities for students to make predictions and for them to feel safe doing so.

Inferring

When students are asked to make an inference, it generally means that they are being asked to explain why something occurs. For example, after placing an inflated balloon in a basin of water, ask students to infer why the balloon floats. Again, it is important to encourage students to take risks when making such inferences. Before explaining scientific phenomena to students, they should be given opportunities to infer for themselves.

Investigating and Experimenting

When investigations and experiments are done in the classroom, planning and recording the process and the results are essential. There are standard guidelines for writing up experiments that can be used with even young students.

- purpose: what we want to find out
- hypothesis: what we think will happen
- materials: what we used
- method: what we did
- results: what we observed
- conclusion: what we found out
- application: how we can use what we learned

Researching

Even at a young age, students can begin to research topics studied in class if they are provided with support and guidelines. Research involves finding, organizing, and presenting information. For best results, teachers should always provide a structure for the research, indicating questions to be answered, as well as a format for conducting the research. Suggestions for research guidelines are presented regularly throughout ***Hands-On Science.***

Using the Design Process

Throughout ***Hands-On Science***, students are given opportunities to use the design process to design and construct objects. There are specific steps in the design process:

1. Identify a need.
2. Create a plan.
3. Develop a product.
4. Communicate the results.

The design process also involves research and experimentation.

Do you or any of your students have a science question you want answered? E-mail your question to Randy Cielen, one of the authors of *Hands-On Science* and a member of the Science Teachers' Association of Manitoba and the National Science Teachers' Association. The address is: books@peguis.com.

Assessment

The *Hands-On Science* Assessment Plan

Hands-On Science provides a variety of assessment tools that enable teachers to build a comprehensive and authentic daily assessment plan for students.

Embedded Assessment

Assess students as they work, by using the questions provided with each activity. These questions promote higher-level thinking skills, active inquiry, problem solving, and decision making. Anecdotal records and observations are examples of embedded assessment:

- anecdotal records: Recording observations during science activities is critical in having an authentic view of a young student's progress. The anecdotal record sheet presented on page 15 provides the teacher with a format for recording individual or group observations.
- individual student observations: During those activities when a teacher wishes to focus more on individual students, individual student observations sheets may be used (page 16). This black line master provides more space for comments and is especially useful during conferencing, interviews, or individual student presentations.

Science Journals

Have the students reflect on their science investigations through the use of science journals. Several specific samples for journalling are included with activities throughout ***Hands-On Science***. Teachers can also use notebooks or the black line master provided on page 17 to encourage students to explain what they did in science, what they learned, what they would like to learn, and how they would illustrate their ideas.

Performance Assessment

Performance assessment is a planned, systematic observation and is based on students actually doing a specific science activity.

- rubrics: To assess students' performance on a specific task, rubrics are used in ***Hands-On Science*** to standardize and streamline scoring. A sample rubric and a black line master for teacher use are included on pages 18 and 19. For any specific activity, the teacher selects five criteria that relate directly to the expectations of students for the specific activity being assessed. Students are then given a check mark point for each criterion accomplished, to determine a rubric score for the assessment from a total of five marks. These rubric scores can then be transferred to the rubric class record on page 20.

Cooperative Skills

In order to assess students' ability to work effectively in a group, teachers must observe the interaction within these groups. A cooperative skills teacher assessment sheet is included on page 21 for teachers to use while conducting such observations.

Student Self-Assessment

It is important to encourage students to reflect on their own learning in science. For this purpose, teachers will find included a student self-assessment sheet on page 22, as well as a cooperative skills self-assessment sheet on page 23. Of course, students will also reflect on their own learning during class discussions and especially through writing in their science journals.

▶

Science Portfolios

Select, with student input, work to include in a science portfolio. This can include activity sheets, research projects, photographs of projects, as well as other written material. Use the portfolio to reflect the student's growth in scientific literacy over the school year. Black line masters are included to organize the portfolio (science portfolio table of contents on page 24 and the science portfolio entry record on page 25).

Note: In each unit of *Hands-On Science*, suggestions for assessment are provided for several lessons. It is important to keep in mind that these are merely suggestions. Teachers are encouraged to use the assessment strategies presented here in a wide variety of ways, and their own valuable experience as educators.

Date: ____________________

Anecdotal Record

Purpose of Observation: ______________________________________

Student/Group	Student/Group
Comments	Comments
Student/Group	Student/Group
Comments	Comments
Student/Group	Student/Group
Comments	Comments

Date: ____________________

Individual Student Observations

Purpose of Observation: ______________________________________

Student ________________________________
Observations

Student ________________________________
Observations

Student ________________________________
Observations

Science Journal

Date: ____________ Name: ____________

Today in science I ____________

I learned ____________

I would like to learn more about ____________

Science Journal

Date: ____________ Name: ____________

Today in science I ____________

I learned ____________

I would like to learn more about ____________

© Peguis Publishers 2000. This page may be reproduced for classroom use.

Sample Rubric

Science Activity: Looking at Seeds

Science Unit: ______

Date: ______

5 - Full Accomplishment
4 - Good Accomplishment
3 - Substantial Accomplishment
2 - Partial Accomplishment
1 - Little Accomplishment

Student	Criteria					Rubric Score /5
	Follows Directions	Displays Curiosity	Makes Detailed Observations	Sorts and Classifies Seeds	Uses Appropriate Vocabulary to Communicate Ideas	
Jarod	✓	—	✓	✓	—	3
Anh	✓	✓	✓	✓	✓	5

Rubric

Science Activity: ____________________

Science Unit: ____________________

Date: ____________________

5 - Full Accomplishment
4 - Good Accomplishment
3 - Substantial Accomplishment
2 - Partial Accomplishment
1 - Little Accomplishment

Student	Criteria					Rubric Score /5

Teacher: ______________________________

Rubric Class Record

Student	Unit/Activity/Date									
	Rubric Scores /5									

Cooperative Skills
Teacher Assessment

Date: ______________________________

Task: ______________________________

____

	Cooperative Skills					
Group Member	Contributes ideas and questions	Respects and accepts contributions of others	Negotiates roles and responsibilities of each group member	Remains focused and encourages others to stay on task	Completes individual commitment to the group	Rubric Score /5

Comments: ______________________________

____

____

____

____

5 - Full Accomplishment
4 - Good Accomplishment
3 - Substantial Accomplishment
2 - Partial Accomplishment
1 - Little Accomplishment

Date: ____________ Name: ____________________

Student Self-Assessment

Looking at My Science Learning

1. What I did in science: ____________________

____

____

2. In science I learned: ____________________

____

____

3. I did very well at: ____________________

____

____

4. I would like to learn more about: ____________________

____

____

5. One thing I like about science is: ____________________

____

____

Note: The student may complete this self-assessment or the teacher can scribe for the student.

Date: ____________ Name: ____________

Cooperative Skills Self-Assessment

Students in my group:

____________ ____________

____________ ____________

Group Work – How Did I Do Today?

Group Work	How I Did (✓)		
	☺	😐	☹
I shared ideas.			
I listened to others.			
I asked questions.			
I encouraged others.			
I helped with the work.			
I stayed on task.			

I did very well in ____________

____

Next time I would like to do better in ____________

____

Date: ____________ Name: ____________________

Science Portfolio Table of Contents

Entry	Date	Selection
1.		
2.		
3.		
4.		
5.		
6.		
7.		
8.		
9.		
10.		
11.		
12.		
13.		
14.		
15.		
16.		
17.		
18.		
19.		
20.		

Date: ____________ Name: ____________________

Science Portfolio Entry Record

This work was chosen by ____________________

This work is ____________________

____

I chose this work because ____________________

____

____

____

____

Note: The student may complete this form or the teacher can scribe for the student.

- -

Date: ____________ Name: ____________________

Science Portfolio Entry Record

This work was chosen by ____________________

This work is ____________________

____

I chose this work because ____________________

____

____

____

____

Note: The student may complete this form or the teacher can scribe for the student.

© Peguis Publishers 2000. This page may be reproduced for classroom use.

Unit 1

Growth and Changes in Plants

Books for Children

Burnett Hodgson, Frances. *The Secret Garden.* New York: Knopf, 1998.

Burns, Diane. *Trees, Leaves and Bark.* Milwaukee: Gareth Stevens Publishing, 1995.

Buscaglia, Leo. *The Fall of Freddie the Leaf.* Austin: Holt, Rinehart & Winston, 1983.

Carle, Eric. *The Tiny Seed.* New York: Crowell, 1970.

Cooper, Jason. *The Earth's Garden: Cactus.* Vero Beach: Rourke Enterprises, 1991.

Delaney, A. *Pearl's First Prize Plant.* New York: HarperCollins, 1997.

Eyvindson, Peter. *Jen and the Great One.* Winnipeg, MB: Pemmican, 1990.

Forsyth, Adrian. *How Monkeys Make Chocolate: Food and Medicines from the Rainforests.* New York: Firefly Books, 1995.

Fowler, Allan. *If It Weren't For Farmers.* Chicago: Children's Press, 1994.

Garner, Alan. *Jack and the Beanstalk.* New York: Doubleday, 1992.

Gibbons, Gail. *Farming.* New York: Holiday House, 1988.

_____. *From Seed to Plant.* New York: Holiday House, 1991.

Hickman, Pamela, and Heather Collins. *A Seed Grows.* Toronto: Kids Can Press, 1997.

_____. *The Kids Canadian Plant Book.* Toronto: Kids Can Press, 1995.

Kerrod, Robin. *Plants in Action.* Bath: Cherrytree Books, 1988.

Krudop, Walter. *Something Is Growing.* New York: Atheneum, 1995.

Landau, Elaine. *Endangered Plants.* New York: Franklin Watts, 1992.

Love, Ann, June Drake, and Pat Cupples. *Canada At Work: Farming.* Toronto: Kids Can Press, 1994.

Lovejoy, Sharon. *Roots, Shoots, Buckets, and Boots: Gardening Together With Children.* New York: Workman Publishing, 1999.

Peteraf, Nancy. *A Plant Called Spot.* New York: Doubleday, 1994.

Pirotta, Saviour. *Trees & Plants in the Rainforest.* Austin: Raintree Steck-Vaughn, 1999.

Primavera, Elise. *Plantpet.* New York: Putnam's, 1994.

Ross, Bill. *Straight From the Bear's Mouth: The Story of Photosynthesis.* New York: Atheneum, 1995.

Stone, Lynn. *The Amazing Rain Forest.* Vero Beach: The Rourke Book Company, 1994.

Suzuki, David. *Looking At Plants.* New York: Wiley, 1991.

Schwartz, David M. *Plant Stems & Roots*. Look Once, Look Again Plants and Animals Science Series. Cypress, CA: Creative Teaching Press, 1998.

_____. *Plant Leaves*. Look Once, Look Again Plants and Animals Science Series. Cypress, CA: Creative Teaching Press, 1998.

_____. *Plant Blossoms*. Look Once, Look Again Plants and Animals Science Series. Cypress, CA: Creative Teaching Press, 1998.

_____. *Plant Fruits & Seeds*. Look Once, Look Again Plants and Animals Science Series. Cypress, CA: Creative Teaching Press, 1998.

Seuss, Dr. *The Lorax*. New York: Random House, 1971.

Creative Teaching Press books are available from Peguis Publishers, Winnipeg

Web Sites

- **http://www.fi.edu**

  Linked to the Franklin Institute. An excellent site for researching "living things" with extensive cross-links and links to various resources. Find sites for classification, adaptation, ecosystems, biomes, habitats, the life cycle, survival, the food chain, and the energy cycle.

- **http://tqjunior.advance.org/3715**

  Learn about the different parts of a plant, their function, and how they grow. Also includes information on how seeds travel and what bees do to help plants.

- **http://socscisvr-01.ucsc.edu/**

  The Life Lab Science Program researches and develops curriculum for elementary school science by using a Living Laboratory school garden. Selected as a "Center of Excellence" by the NSTA and the NSF.

- **http://aggie-hoticulture.tamu.edu/ kindergarden/ index.html**

  Texas Horticulture Program: go to "just for kids," an excellent introduction to the many ways children can interact with plants and the outdoors. Includes information on the nutritional value of plants, human issues in horticulture, and composting for kids.

- **http://greenschools.ca/seeds**

  Society, Environment, and Energy Development site. Become an Environmental Green School and join the 4500 schools across Canada that are currently registered in this program (also called Learners Action).

- **http://www.owu.edu/~mggrote/pp/**

  Project Primary is a collaboration between elementary school teachers and professors from different departments. Find hands-on activities for teaching botany.

- **http://www.garden.org/home.html**

  American National Gardening Association. This site has a "kids and classrooms" link to "Growing Ideas – A Journal of Garden-Based Learning," written for and by teachers. Articles include "Celebrating Seeds," "Food Plant Life Stories," "Sparking Student Inquiry," and much more.

- **http://www.urbanext.uiuc.edu/**

  Urban Programs Resource Network (University of Illinois): click on "just for kids" and go to "The Great Plant Escape." Each case study of plant life gives clues to students to solve the mystery. Includes Teacher Guide.

- **http://www.web.net/~greentea/**

  Green Teacher Magazine is written by and for teachers committed to environmental and global education. Includes online articles such as "Re-mystifying the City," "Waste Reduction," "The Outdoor Classroom," and much more. *Green Teacher* also offers extensive links.

- **http://www.kortright.org**

  Located just ten minutes away from Metro Toronto, The Kortright Centre is Canada's largest environmental education centre – with over sixty different programs that have been designed to complement classroom curriculum.

Introduction

This unit focuses on the study of plants, specifically the characteristics and requirements of plants and their patterns of growth. Students will observe and investigate plants in their local environment. They will also learn about the importance of plants for humans and animals.

Students will learn about the similarities and differences in the physical characteristics of different plant species and the changes that take place in different plants as they grow.

Students will also investigate the effects of changes in environmental conditions on plants, find out why plants are important to other living things, and learn about the effects of human activities on plants.

Ensure that you have collected live plants for the classroom, as well as an assortment of seeds and bulbs. When choosing plants, consider variety and durability. Bean plants grow well, and the students will be able to observe them develop from seed to adult plant in a comparatively short period of time. Also ensure that the plants you select display a variety of different leaf patterns, stems, and root systems. You will also need soil and various containers in which to grow these plants.

You may wish to set up a plant centre in a sunny area of the classroom so that the students will have one location in which to observe, record, and discuss changes on a regular basis. This type of centre encourages the students to make predictions about future changes in the plants, and to infer the reasons for such changes.

It is suggested that you collect numerous pictures of plants for use in activities and at centres. Resources for collecting pictures include:

- old calendars
- seed catalogues
- gardening magazines
- horticultural societies
- departments of forestry, natural resources
- forestry and environmental associations
- photographs from home

Contact government departments and associations well in advance of studying the unit. You may be able to obtain other related materials and services such as booklets, posters, films, and presentations for classroom use.

It is not necessary to teach this unit, or any other, in one block. As outdoor plant life is abundant during the fall and spring, you might consider splitting up the activities to focus on plants during these seasons, or study plant life on an ongoing basis throughout the year.

Science Vocabulary

Throughout this unit, teachers should use, and encourage students to use, vocabulary such as: *basic needs, light, water, air, space, growing medium, roots, stem, leaves, flowers, seeds, pistil, stamen, life cycle, fibrous root, tap root, deciduous, coniferous, nutrients,* and *soil.*

Materials Required for the Unit

Classroom: soft paintbrushes, chart paper, felt pens, water, masking tape, plant journal (included), graph paper (included), Hula-Hoops (or string), construction paper or Manila tag, scissors, glue, grid paper, pencils, crayons, markers, smocks or paint shirts, oil pastels, metric ruler, permanent marker, white art paper, research outline (included), 30-cm rulers, clipboards, labels

▶

Books, Pictures, and Illustrations: newspaper, newspaper flyers, magazines, pictures of harmful plants (e.g., poison ivy and stinging nettle), *Straight from the Bear's Mouth: The Story of Photosynthesis* (a book by Bill Ross), *Ready, Set, Grow: A Guide to Gardening with Children* (a book by Suzanne Frutig Bales), books on uses of plants, pictures of various trees, shrubs, and plants (from seed catalogues, calendars), pictures of plants through the seasons

Household: glasses, food colouring, knife, spray mister, paper plates, white sheet, pans (for boiling water), potato masher, fork, measuring cups, containers (for dyeing cloth), aluminum baking pans (one containing soil and one containing sod), toothpicks, containers (for water)

Equipment: camera (optional)

Other: small plants in flower, lunch-size paper bags, rope or pylon markers, plant samples (roots, stems, leaves, flowers, and seeds), samples of fibrous and tap roots, potted plants, celery, white carnations, cactus, variety of leaves, bean seeds, potting soil, (e.g., vermiculite, peat moss, heavy clay, sand, gravel), planters, bean plants, radish seeds, pots for planting seeds, small stones, baskets (filled with fresh fruits, vegetables, spices, and herbs), heat source, Popsicle sticks, water bucket, thick wood block (wedge), plant nutrients (fertilizer), vegetable seeds, plants for making dye (e.g., berries, beets, carrots, purple grapes)

1 Parts of a Plant

Science Background Information for Teachers

The basic parts of plants include: root, stem, leaves, flower, pistil, stamen, sepal, and seeds.

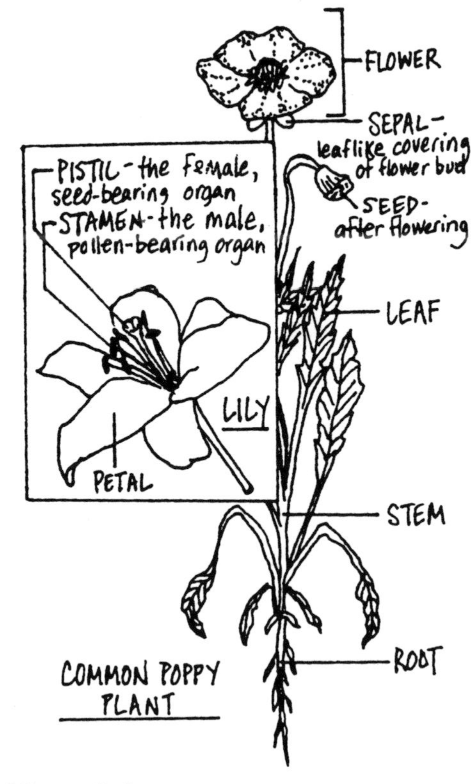

Materials

- newspaper
- small plants in flower, one per student or per pair of students (buy cell packs at a garden centre)
- soft paintbrushes
- chart paper, felt pens

Activity

As an introduction to plants, have students brainstorm a list of all the plants they know. Record their ideas on chart paper, and discuss what all of these plants have in common.

Have students work individually or in pairs. Give each student or pair of students some newspaper, a paintbrush, and a flowering plant. Ask the students to cover their desks with the newspaper, then gently remove the plant from its pot and lay the plant on top of the newspaper.

Have the students use the paintbrush to carefully brush away as much soil as they can. Then have them identify any parts of the plant they know. Ask:

- What do these plants all have in common?

As the students identify the parts – including roots, stem, leaves, flowers, and seeds – write the name of each part of the plant on chart paper. Have the students predict and discuss the function of each part.

Once all of the parts of the plant have been identified, have students sketch their plant on their activity sheet and label the diagram.

Activity Sheet

Directions to students:

Draw a diagram of your plant. Label these parts: root, stem, leaf, flower, seeds (1.1.1).

Note: Additional plant parts can also be labelled if observed on the plant (e.g., pistil, stamen, sepal).

Extension

Plan and implement a language arts novel study, using *The Secret Garden* by Frances Hodgson Burnett.

Assessment Suggestion

Through individual conferences, or by reviewing the students' activity sheets, determine if students can identify and describe the basic functions of the different parts of the plant. Use the rubric on page 19 to identify criteria and record results.

Date: ____________________ Name: ________________________________

Basic Parts of a Plant

Label the following on your diagram: root, stem, leaf, flower, seeds

2 Classifying Plants

Materials

- lunch-size paper bags (one per student)
- rope or pylon markers (to identify areas where students may collect materials)
- chart paper, felt pens
- pictures of harmful plants, such as poison ivy and stinging nettle
- samples of plant roots, stems, leaves, flowers, and seeds (other than those that will be easily collected on the nature walk)

Activity

Note: Identify an area where students can collect examples of plant life such as leaves, bark, twigs, pine cones, dandelions, and so on. Ensure that it is safe and acceptable to collect these items in the designated area. (If you are going to use a municipal or provincial park, get prior approval and follow any recommended guidelines.)

To start this activity, discuss the importance of collecting plants in a safe way. Ask students:

- Are you aware of any types of plants that are harmful to humans? (Some examples are poison ivy, poison oak, stinging nettle, poinsettia.)
- Why are they harmful? (Some examples are rashes, harmful if eaten.)
- How can you be safe from harmful plants?

Display the pictures of harmful plants for students to examine and describe. Encourage them to learn to identify these plants so that they will stay away from them.

When you reach the site of your nature walk, tell students that they are going to collect different parts of plants on their walk. Explain that they will be comparing and sorting what they find when they return to the classroom. Encourage them to look for various samples of different parts of plants (e.g., root, stem, leaf, flower, seed). Whatever they collect must be able to fit into their paper lunch bag.

Designate, with ropes or pylons, the area where students are going to be collecting. Remind the students that they are allowed only to collect plants from this area. Also encourage them to be respectful of living things; for example, pick up leaves and bark from the ground rather than pick them off a tree.

When students return to the classroom, have them examine what they found. Have them select a rule as to how they will sort their materials (e.g., by part, colour, shape). After students have sorted their own materials, have them walk around the room to observe how other students sorted their items. Ask questions such as:

- Are the leaves pointed or rounded?
- Is the bark smooth or rough?
- Are the seeds the same size, colour, or weight?
- What colour are the flower petals?

Encourage students to classify the plant parts in several different ways and then challenge other students to identify the sorting rules.

During these discussions, encourage students to closely observe the plant parts and describe them in detail, using all of the senses (except taste). Also compare their collected plant parts with those you have provided for the activity.

Activity Sheet

Note: Students will need to have examples of roots, stems, leaves, flowers, and seeds before they can complete the activity sheet. If not enough samples were collected on the nature walk, you may need to have extra on hand for students to observe and examine.

Directions to students:

Select two samples of each plant part to compare. Complete the classification chart. You may wish to include a diagram with your description (1.2.1).

▶

2

Extensions

- Have students play a matching game with the plant parts they have found. Have a student ask a question such as, "Who has a smooth piece of bark that is white?" Students who have a match may hold up their matching plant part.
- Send students on a scavenger hunt for specific items (e.g., the largest leaf they can find, the plant with the most number of seeds).
- Have students identify the names of plants based on the characteristics of their parts (e.g., a tree with smooth white bark is a birch tree).
- Create a class "word splash." Have students brainstorm a list of words associated with plants. Record their suggestions randomly on chart paper. Sort the words (by parts, types, and so on) by circling them with different colours of marker. As an alternate activity, focus on specific concepts; for example, with a blue marker, underline all plants than can be used for food.

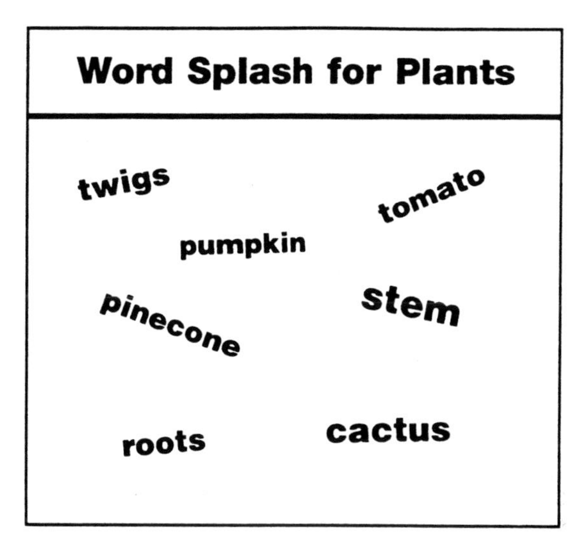

Activity Centre

Display nonfiction books about different types of plants. Set up a display of the different plant parts that the students collected on their nature walk. Encourage students to use the books at the activity centre to identify the plants based on their characteristics.

Date: ____________ Name: ____________________

Plant Classification Chart

Plant Part 1	Plant Part 2	How They Are the Same
root	root	
stem	stem	
leaf	leaf	
flower	flower	
seed	seed	

3 Special Features of Plants

Science Background Information for Teachers

Roots: Roots anchor a plant in place, and seek out and store moisture and nutrients for the plants. The two types of root systems are tap roots and fibrous roots. A tap root is a long, thick root that grows down deep and straight (e.g., dandelion root, carrot, turnip). A fibrous root has many root tips that spread out in all directions (e.g., grass, most potted plants).

Stem: The xylem cells of the plant stem transport water and other nutrients from the roots to the leaves. The stem also carries the food made in the leaves to all parts of the plant.

Leaves: Leaves make food for the plant to live and grow. Leaves are filled with chlorophyll (this gives the leaves their green colour). Chlorophyll helps turn sunlight, water, minerals, and air into food for the plant.

Materials

- paper lunch bags
- clipboards
- pencils
- trowels
- newspaper
- samples of fibrous and tap roots
- potted plants (optional – see Note: Activity A)
- glasses of water (two per group)
- food colouring
- fresh pieces of celery with tops (one per group)
- white carnations (one per group)
- knife (for adult use only)
- cactus
- variety of leaves collected by students
- *Straight from the Bear's Mouth: The Story of Photosynthesis*, a book by Bill Ross
- water

As a class, review the parts of the plant. Now explain to the students that they are going to conduct a number of experiments to investigate the special features of plants. Through these experiments, they are going to learn how each part of the plant has a special role in the survival of that plant.

Divide the class into small groups (approximately four students per group). Keep students in these groups as they work through each activity.

Activity: Part One: Root Systems

Note: For this activity, have the students collect their own root samples from weeds and other plants in the neighbourhood. If this is not possible, you can use a variety of potted plants to investigate root systems.

Ask the students to predict and explain the job of the roots of a plant. Have students fill in the prediction box at the top of activity sheet A (1.3.1).

If the season permits, take the students for a walk in the schoolyard or in a local park. Have students work in their small groups. Give each group a paper lunch bag, clipboards, pencils, and activity sheet A. Ask the students to find a weed in the ground. Before they dig up the weed, have the students draw what they think the root system looks like. Once they have drawn their prediction, ask the students to dig up their weed carefully and place it in their paper lunch bag. Once you are back in the classroom, have the students soak their weed in water to remove any soil around the weed. Ask students to compare their prediction of what they thought the roots would look like with the actual root system. Have them draw the actual root system next to their prediction.

Display the weeds on a table covered with newspaper. Once the students have had an opportunity to view the root systems of different weeds, ask:

- Are all root systems the same?
- How are they different?
- Was it easy to dig up your root from the ground?
- What special role do you think roots play in the survival of a plant?

Introduce the terms *tap root* and *fibrous root*. Have the students sort their weeds according to the type of root system.

Note: If students did not collect examples of both fibrous and tap roots, use samples that you have provided to observe, compare, contrast, and discuss.

Activity Sheet A

Directions to students:

1. Fill in your prediction at the top of the page.
2. Draw your prediction of the root system of the weed you have found.
3. Draw the actual root system of the weed you have found.
4. In your own words, describe the special function of roots.

Extension

Grow plants from vegetables (e.g., a potato and a carrot) and from cuttings. Observe the different root systems. Compare the growth of these plants.

Activity: Part Two: Stem

Ask students to predict and explain the job of stems in a plant. Have students fill in the prediction box at the top of activity sheet B (1.3.2) before starting the activity.

Provide the following instructions to each group:

1. Fill two glasses with water.
2. Put a few drops of food colouring in each glass.
3. Put the celery stalk in one glass and the carnation in the other.
4. Draw the plants on your activity sheet (initial observation).
5. Set the glasses in the sunlight. Leave them overnight.
6. Record your observations on your activity sheet.

Ask the students:

- Can you explain the role of the stem in the survival of a plant?
- Why is this an important role?

Using a sharp knife (adult only), cut off a section of the stems of the celery stalk and carnations so that students can examine the inside of the stem. Ask:

- What does the inside of the stem look like?
- Can you locate the tubes that carry the water up the stem in the celery and in the carnation?
- How did you locate these tubes?
- What else do you think stems may carry? (They carry nutrients to different parts of the plant.)

Display the cactus and have students identify it and locate the stem. Explain to the students that cacti have special stems called fleshy stems. Ask:

- Do you think the role of the stem in a cactus is different from any other plants?
- What is the habitat of a cactus like?
- Why is the stem so important for a cactus?

Activity Sheet B

Directions to students:

1. Fill in the prediction box at the top of the page.
2. Draw a diagram of the celery and the carnation in the coloured water at the beginning of the experiment.
3. Draw your observations of the celery and carnation stems at the end of the experiment.
4. In your own words, explain the special function of the stem.

Extension

Vary this experiment by using more than one colour of food colouring and by splitting the stems of a variety of flowers (e.g., white carnations, Shasta daisies, and white roses). Try splitting stems into halves, thirds, or quarters and putting each stem section into a different colour of water for several hours or overnight.

Activity: Part Three: Leaves

Note: The concept of photosynthesis can be quite abstract for students at this age: it is impossible for students to observe directly the production of food by the plant. The activities that follow should assist students in gaining a basic understanding of the importance of leaves.

Provide each group with several leaves (from both coniferous and deciduous trees) to observe, examine, and discuss. Encourage them to identify similarities and differences. Ask students to fill in their prediction box at the top of activity sheet C (1.3.3). You can also have them draw diagrams of five different types of leaves on the back of their activity sheets.

Read the story *Straight from the Bear's Mouth: The Story of Photosynthesis* by Bill Ross.

Note: You may wish to read only those parts from the book that are appropriate for this investigation.

Following the story, ask the students:

- Why are leaves green?
- Why does the plant need the green part (chlorophyll)?
- What else does a green leaf need in order to make food for the plant?

Activity Sheet C

Directions to students:

1. Fill in the prediction box at the top of the page.
2. Draw all of the things a green leaf needs in order to make food for the plant (recipe for plant food).
3. In your own words, explain the special function of leaves.

Extensions

- Provide three potted, newly sprouted plants of the same size. Label the plants (1, 2, and 3). Pinch all the leaves off Plant 1. Pinch half the leaves off Plant 2. Do not pinch any leaves off Plant 3. Care for the plants in the same way, making sure they receive the same amount of sunlight and water. Have students observe the plants and measure the differences in growth. Discuss the importance of leaves in the growth of the plant.

- **Leaf Rubbings:** Place a leaf under drawing paper and lightly colour the paper with a crayon. The stem and leaf pattern will be clearly displayed.

- Leaves can be saved for further study or for classroom display by placing them between sheets of wax paper, covering the wax paper with a cloth, and pressing with

a warm iron. Leaf patterns will be apparent and the leaves will be permanently maintained.

- **Leaf Skeletons:** Collect various leaves and place them between sheets of newspaper. Place books on top of the newspaper to add weight. After a few days, when the leaves have become dry and brittle, place one leaf between two sheets of paper and use a hammer to gently pound the entire leaf surface. Remove the top sheet of paper and observe the leaf skeleton. This is an excellent way to closely examine veins, stems, and the framework of leaves.
- **Breathing Leaves:** Using one plant with several leaves, rub half the leaves on both sides with Vaseline. Tie a string or a twist tie loosely around the stem of each coated leaf so that you will remember which leaves have been treated. Leave the plant for several days, observing the changes in the leaves; the leaves covered with Vaseline will begin to die because they cannot take in air.
- **Patterned Leaves:** Using plants that have several large leaves, paper clip small cutout circles or rectangles of black construction paper over sections of some of the leaves. Place the plants in a sunny area for several days, then remove the construction paper and observe the leaves. The areas covered with paper will be much paler, because these areas did not receive the sunlight required to produce chlorophyll, which gives the plants their green colour. (Sunlight is necessary for photosynthesis, a process in which the plant creates its own sugar for food.)
- **Moisture in Leaves:** Using a plant with several leaves, place a small plastic bag over a few of the leaves, then tie the bag securely with string. After a few days you will notice water droplets in the bag because plants that use water also disperse water into the air.
- Use leaves for printmaking by dipping them in paint and pressing them onto art paper.
- Gather a group of leaves and have students sort them according to size, veins, edges, texture, colour, and shape.

Assessment Suggestion

Through individual conferences with students, have the students identify the parts and explain the functions of a displayed sample plant. Use the individual student observations sheet on page 16 to record results.

Date: ______________ Name: ______________________

Roots

Prediction Box

I think the function of the root is ______________

____

____

Predicted Root System **Actual Root System**

Function of Roots

The function of roots is ______________

____

____

Date: ____________ Name: ____________________

The Stem

Prediction Box

I think the function of the stem is ______________

____

____

First Observations:

Final Observations:

Function of the Stem

The function of the stem is ______________

____

____

Date: ______________ Name: ________________________

The Leaves

Prediction Box

I think the function of leaves is ____________________

____

____

Recipe for Plant Food

Function of Leaves

The function of the leaves is ____________________

____

____

4 Life Cycle of a Plant

Science Background Information for Teachers

The stages of the life cycle of a flowering plant are:

1. seed
2. seedling
3. adult plant
4. adult plant with flowers
5. adult plant with fruit (back to step 1)

Note: Many flowering plants do not really bear fruit, at least in the "typical" definition of fruit (e.g., apple, orange). The seed of the plant, however, is scientifically referred to as fruit while still in the flower. An example of this is wheat. The wheat grains are considered to be the "fruit of the plant." In the case of a bean plant, the actual bean is considered the fruit.

Materials

- bean seeds
- potting soil
- planters or pots
- water
- camera (optional)
- mister (to water plants)

Note: If you are using a recycled spray bottle, make sure the ingredients that were in it were nontoxic.

Activity

Explain to the students that throughout their plant unit, they are going to track the growth of plants as the plants go through their life cycle. Ask:

- What does a plant need in order to survive? (Plants need water, sunlight, soil, and warmth.)

Divide the class into working groups. Have the students observe and examine the bean seeds. Encourage the students to describe the seeds.

Soak the seeds overnight in water. Have students draw and label the first stage of the cycle for their plant on their activity sheet.

Have each group plant some seeds in a planter. Place the planter in a warm, sunny area. Have the groups take care of their plant by watering it as needed. You may want to have the students use a mister to water the plant so that it is not overwatered.

As the plants begin to grow, have the students label and illustrate the life cycle of the plant on their activity sheet.

Talk to the students about what they observe as the plants grow. Ask questions such as:

- When did the leaves start to appear?
- What colour were the leaves?
- What shape were the leaves?
- Did the leaves appear in pairs or one by one?
- What is the stem like?
- What parts of the plant are changing as it grows?
- What parts of the plant are staying the same as it grows?

You may wish to take pictures of the plants and document all of the stages of development.

Note: Plants may need to be transplanted into larger pots once they have grown.

Activity Sheet

Directions to students:

Label the stages of the plant's life cycle. Be sure to draw a diagram of your plant at each stage (1.4.1).

Extensions

- Have students measure, record, and graph the growth of their plants. Have groups compare the growth of their plants with others.
- Provide Plasticine for the students to make models of their plants at different stages of the life cycle.
- Plant bulbs to observe another way that plants can be grown. Compare the life cycle of plants from bulbs to plants grown from seeds.
- Read different versions of *Jack and the Beanstalk*. Compare the story to the students' knowledge of how bean plants grow.

Activity Centre

Once the plant has gone through the stages of the life cycle, develop the photographs. Mount the pictures on poster board and laminate them. Place the pictures in an envelope at the activity centre. Have the students sequence the pictures, depicting the stages of the plant's life cycle.

Date: ____________________ Name: ________________________________

Life Cycle of a Flowering Plant

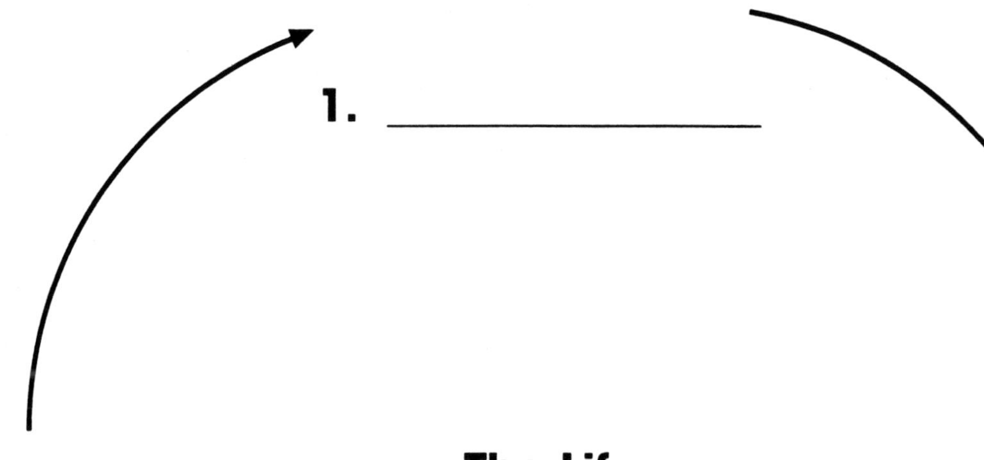

1. ____________________

The Life Cycle of a Flowering Plant

5. ____________________

2. ____________________

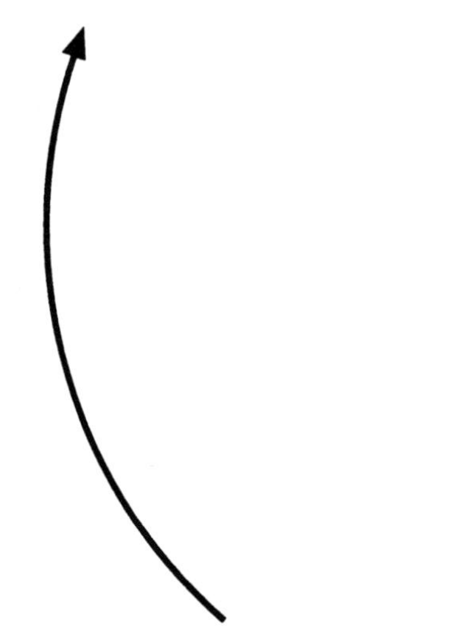

4. ____________________

3. ____________________

5 Plants Need Water and Light

Materials

- 4 bean plants in containers of soil (or other plants of the same type and size)
- 30-cm rulers
- containers of water
- masking tape
- felt pens
- 4 charts for recording results made prior to beginning this activity (included as activity sheets)
- graph paper (included)

Activity

Display the plants for the students to examine. Ask:

- What does a plant need to live?
- How can you properly care for plants?
- What do you think would happen to a plant if it did not have water?
- What do you think would happen to a plant if it did not have light?

Have the students test their predictions. First have them observe and describe the present states of all four plants, including their heights (which should be measured and recorded on the charts).

Stick masking tape on each pot and label the plants as follows: Control Plant, Plant With No Sunlight, Plant With No Water, and Plant With No Sunlight or Water. Place the plants in appropriate locations according to their labels.

Observe the plants on a regular basis. Note changes, measure their heights, and record this information on the charts.

Discuss changes as they occur, comparing the control plant to the plants with no sunlight, no water, and no sunlight or water.

Activity Sheet

Directions to students:

Use the charts to record the results of your experiments with plants (1.5.1-1.5.4). Then use graph paper to make a bar graph of your results (1.5.5).

Note: Depending on the background experience of your students in constructing graphs, you may choose to guide this activity to ensure that they include all required information on the graph (see pages 9-10). This may also be done as a whole-class activity on large graph paper.

Extensions

- Write up the investigations from the activity, using the scientific method outlined on the extension activity sheet (1.5.6).
- Conduct an experiment to show that the direction a plant will grow depends on its source of light. Plant a sprouting potato in soil. Place it at one end of a shoebox set on its long side (as shown) and then cut a hole in the other end of the box.

  Tape dividers inside the box in two different positions, making sure that they do not extend the full width of the box. Place the lid on the box. Place the box in sunlight, with the hole directed toward the Sun. The potato shoot will find its way through the maze toward the light source. Since sunlight is needed to produce chlorophyll, the shoot will not be green. The students will be able to observe the tip of the shoot turning green as it grows toward the sunlight.

Date: ____________________ Name: ________________________________

Control Plant

Date	Height	Observations

Date: ____________ Name: ____________________

Plant With No Water

Date	Height	Observations

Date: ____________________ Name: ______________________________

Plant With No Sunlight

Date	Height	Observations

Date: ____________________ **Name:** ______________________________

Plant With No Water or Sunlight

Date	Height	Observations

Date: ____________________ **Name:** ________________________________

Date: ________________ Name: ____________________________

Experimenting With Plants

Purpose (what you want to find out): ______________

____

____

Hypothesis (what you think will happen): __________

____

____

Materials (what you will need): __________________

____

____

Method (what you did): ________________________

____

____

____

____

____

____

Results (what happened): ______________________

____

____

Conclusion (what you learned): __________________

____

____

Application (how you can use what you learned):

____

____

6 The Effects of Soil and Nutrients on Plants

Materials

- 3 different types of soil, such as good potting soil, vermiculite, peat moss, heavy clay, sand, gravel, and so on
- plant nutrient (fertilizer)
- radish or bean seeds (or other relatively fast-growing seeds)
- pots for planting seeds
- small stones
- water
- masking tape
- felt pens
- chart paper
- plant journal (activity sheet) (1.6.1) (Make 6 copies of the sheet for each student to begin with. More sheets may be required as the experiment progresses.)

Activity

Display the seeds for the students to examine. Ask the students:

- What do these seeds look like?
- How could you find out what kind of plant these seeds will grow into?
- What should you do to the seeds before you plant them?

Soak the seeds in water overnight.

Review procedures for plant care by asking:

- What do plants need to grow?
- What do plants need to grow in?
- Do you think the type of soil will make a difference to the way the seeds grow?
- What could you add to the soil that might make a difference to the way the seeds grow?

Have the students examine the types of soil. Ask:

- How are the soils different?
- In which soil do you think the seeds would grow the best?
- In which soil do you think the seeds would not grow well?

Have the students test their predictions by doing the following experiment.

Have each group place a few stones in the bottom of six pots for drainage. Plant seeds in the various soil types. Label the pots according to the type of soil used and whether or not a nutrient supplement was used.

Sample labels:

- Potting Soil with Fertilizer
- Potting Soil without Fertilizer
- Peat Moss with Fertilizer
- Peat Moss without Fertilizer
- Sand with Fertilizer
- Sand without Fertilizer

Review plant location and other factors that affect growth, such as water supply and access to sunlight. Have the students choose a good location in the classroom for the plants.

While the plants are growing, observe them every few days. Measure their growth and other changes and record this information on chart paper.

Have the students print the name of the soil type at the top of each plant journal sheet, and indicate whether a nutrient supplement was added to the soil. On a regular basis, have the students observe the plants, and record the date as well as their observations about the growth of the plants.

After the experiment, have the students decide which soils are best to grow seeds in, and whether or not the nutrients had an effect on the growth of the plants. You may also wish to have students order the soil types (including plants with nutrients) from worst to best.

6

Activity Sheet

Note: Recording observations is an important aspect of experimentation. Explain this to the students and encourage them to observe carefully and record accurately.

Directions to students:

Use your plant journal to record your observations about how the plants grow in different types of soil (1.6.1).

Extensions

- Have students graph the growth of the plants. Ask students questions about the growth of the plants to see if they can interpret the graph.
- Challenge the students to use the design process to construct an environment that enhances plant growth. Encourage them to consider the basic needs of plants and factors that enhance plant growth (moisture, sunlight, warmth, soil, nutrients, and so on). Examples might include designing and constructing a self-watering planter or a plant warmer.
- Have students design and construct terrariums to determine how these containers enhance plant growth.
- Invite a guest speaker from a local garden centre. Have the speaker discuss the different needs of various types of plants, the different types of soils and fertilizers available.

Assessment Suggestion

On a regular basis, review students' observations in their plant journals. Check for careful observations and accurate recording. Use the anecdotal record sheet on page 15 to comment on students' use of the plant journals.

Date: ______________ Name: ______________________

Plant Journal

Type of Soil:______________ Added Nutrients: Yes ☐ No ☐

Date	Observations

7 Plants and Seasonal Changes

Materials

- pictures of various trees, shrubs, and plants (from seed catalogues, calendars, and so on)
- 4 pictures showing plants throughout the seasons (1.7.1-1.7.4)

Note: For teaching purposes, you may want to colour in pictures that show plants throughout the seasons.

- paper and pencils for recording data on walks (Consider using notepads or paper on clipboards, or attach paper to firm sheets of cardboard so that the students have sturdy surfaces to write on.)
- camera (optional)

Activity

Display pictures of trees, shrubs, and small plants for the students to observe and examine. Ask the students:

- How are these plants different?
- How are these plants the same?
- What do you think happens to each plant in winter?
- How do the plants change when spring comes?
- What do the plants look like in the summer?
- What happens to plants in the fall?

After this introductory activity, take the students for a walk to a nearby park or through the neighbourhood. Have the students observe the various plants, record the names of those that they know, and give descriptions of all plants.

Note: This activity will vary depending on the season. Even in the middle of winter, some plant life will be apparent. Consider taking nature walks several times during the school year to observe changes in plant life during the seasons.

As each plant is observed, stop and discuss it. Ask:

- What does the plant look like now?
- How will it change with the next season?
- Will the plant die?
- Will it live through the seasons?
- If it lives, how will it change throughout the seasons?
- Will it stay exactly the same?

Have the students predict which plants will die in the fall, which trees will lose their leaves, and which will remain in generally the same state year round.

If a camera is available, take photographs of various plants. Use the pictures to further examine the plants in the classroom, and for making comparisons during the subsequent season.

In the classroom, continue the plant discussion. Classify trees into coniferous and deciduous groups. Other plants can be classified according to the ones that die in winter (annuals) and the ones that lie dormant and revive again in spring (perennials).

Display the four pictures showing plants throughout the seasons. Discuss the observable changes in these plants as a result of seasonal changes. Classify the plants in the pictures according to the ones that change, the ones that do not change, and the ones that die.

Activity Sheet

Directions to students:

Look at the pictures of trees in summer. Draw pictures to show what these trees will look like in the winter (1.7.5).

▶

7

Extensions

- As a class, "adopt" a tree to observe throughout the seasons. Provide opportunities for students to observe, diagram, and write about the tree on a regular basis.
- In the fall, look at the flowering plants to see how their seeds will be dispersed. Save these seeds to plant outside in the spring, or plant them indoors at any time.
- Observe new plant growth in the spring, noting the buds on the trees, shoots in the ground, and so on.
- Read the story, *The Fall of Freddie the Leaf*, by Leo Buscaglia. Have students illustrate the life cycle of the tree in the story through the different seasons.

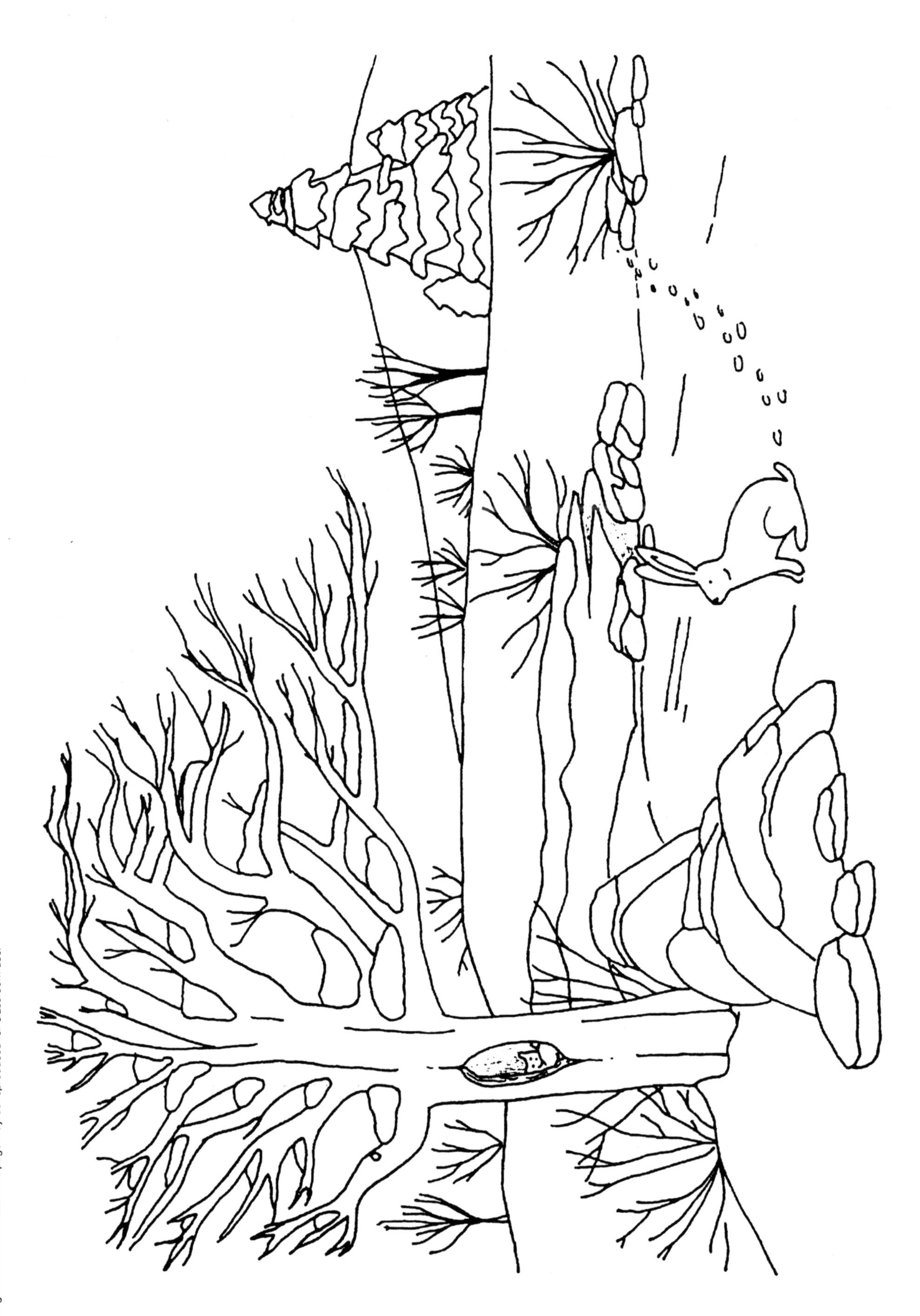

Date: ____________ **Name:** ____________________

Trees in Summer and in Winter

<table>
<tr><td>
Fir tree in summer</td><td>Fir tree in winter</td></tr>
<tr><td>
Oak tree in summer</td><td>Oak tree in winter</td></tr>
</table>

8 Plants That We Eat

Materials

- basket filled with fresh fruits, vegetables, spices, and herbs (include something from each plant part: root, stem, flower, leaf, seed, and fruit; also include a few unique foods from plants, such as chili peppers, leeks, and shitaki mushrooms)
- Hula-Hoops or string (tied to make a circle)
- labels (Print the name of a different plant part - stem, roots, leaves, fruit, flower, seeds - on large index cards.)
- knife (optional)
- paper plates (optional)
- grocery flyers, food magazines
- white art paper
- construction paper or Manila tag
- scissors, glue

Activity

Have students sit in a circle. Place the basket of plant foods in the centre of the circle. Place the Hula-Hoops or string circles around the basket. Tell the students that the basket is filled with different types of fruits, vegetables, herbs, and spices that we eat. Hold each item up and have the students identify the item.

As a group, read the plant part labels and place each one in the middle of a Hula-Hoop. Hold up each item once again and ask the students:

- Which part of the plant does this come from?

Sort each item into the appropriate group.

You may wish to cut up the different food items so that students can sample them.

Note: Make sure you are aware of any student allergies before students touch or sample food.

Provide each student with a piece of art paper. Have students divide their sheet into six sections and label these sections Fruit, Roots, Leaves, Stems, Flowers, Seeds.

Have the students draw or cut out from flyers and magazines examples of foods that come from plants. Instruct them to sort their pictures into each of the six groups, then glue the illustrations into the appropriate sections of the art paper. Suggest that they have at least two examples in each group.

Extensions

- Introduce the students to a variety of foods from plants that are common in other countries, such as plantain, mango, sugar cane, and so on. Extend the activity by sorting these foods according to which part of the plant each comes from.
- Have pictures of different types of food from various parts of the plant on individual cards. Have students play a game of concentration. If they match two foods that come from the same part of the plant, it is considered a match (e.g., a carrot and radish are both from the root of the plant).

Assessment Suggestion

Have the students create a menu that uses various types of food from different parts of plants. Challenge students to calculate the cost of the meal, using prices from grocery store flyers. Use the individual student observations sheet on page 16 to record results.

Note: If possible, as a class, prepare some of the meals that students have designed.

9 Where Does Your Garden Grow?

Materials

- grid paper (3 m x 2 m)
- packages of vegetable seeds
- pencils, pencil crayons, markers
- chart paper
- *Ready, Set, Grow: A Guide to Gardening with Children*, a book by Suzanne Frutig Bales

Activity

Select relevant pages to read from the book *Ready, Set, Grow: A Guide to Gardening with Children*. Ask the students:

- What did you learn about gardening?
- How do humans plan their gardens?
- What do you have to remember when planning your garden?
- This story talked about growing plants in a garden. Where else are plant crops grown?
- Which crops grow on a farm?
- Which plants grow in an orchard?
- Do you know some places in our local community in which plant crops are grown?
- Do any of you have a garden at home?
- What type of plants do you grow in your garden?

Divide chart paper into three columns: Farms, Orchards, and Home Gardens. As a class, record a list of possible types of crops grown in each of these areas.

Divide the class into groups of three or four students. Give each group several types of vegetable seeds. Have the students read the back of the packages to find out specific information about growing the seeds.

Have the members of each group come to a consensus when identifying the vegetables that they would like to grow in their own garden. Give each group an activity sheet to plan a rough draft of its garden, and map the garden out. Explain to the students that each garden map must include:

- Labels identifying the different types of vegetables to be grown.
- Pictures of where the vegetables will be grown and how they will be spread out.
- A note under each label identifying how long it will take each vegetable to grow and any special care required.

When students have completed their rough draft, have them record the final garden map on large grid paper.

Once the students have finished their planning, have them share their vegetable garden plans with the class. Ask questions, such as:

- How did you reach a consensus about what you wanted to plant in your garden?
- Explain where you planted your seeds and why.
- Why is it important to plant the seeds far enough apart?
- Why is it important to make a map of your garden before you plant it?
- How would you look after your garden?

Activity Sheet

Directions to students:

In your groups, plan your garden (1.9.1).

Extensions

- Agree on one garden plan and plant a vegetable garden outside in the spring. Assign duties on a rotational basis. Have each group take care of the garden as it grows.
- Invite a local farmer or gardener to the classroom to speak to students about growing plants.

- Visit a local orchard or farm. This will enable students to observe firsthand how plants are grown.

Activity Centre

Provide mural paper and art materials for the students to make a garden mural. Brainstorm a list of things that might be found in a garden (e.g., plants, seeds, insects, animals, gardening tools, farmer, gardener, scarecrow). Post the list at the centre.

Assessment Suggestions

- Use the cooperative skills teacher assessment sheet on page 21 to record students' ability to work together.
- Have students complete a cooperative skills self-assessment sheet on page 23 to reflect on their participation in their groups.

Names: ____________ ____________ ____________ ____________

Date: ____________

Plans for Our Garden

Vegetables to be planted ____________ ____________ ____________

____________ ____________ ____________

____________ ____________ ____________

10 Plants and Their Importance to Humans

Materials

- chart paper, felt pens
- catalogues, magazines, and newspapers
- glue
- scissors
- mural paper
- books on uses of plants (from local and school libraries)
- research outline (included) (1.10.1)
- resources on plants (e.g., CDs, internet, brochures, books)

Activity A

As a class, discuss the variety of useful products that are made from plants. (Answers may include: foods, clothing, lumber, medicines, dyes, paper, and so on.) On chart paper, record a list of products made from plants.

Divide the class into working groups. Have students work together in their groups to make a collage, using pictures from catalogues, magazines, and newspapers, of products made from plants.

Display the collages in the classroom.

Activity B

Have each student select one item from the list on the chart paper to research. Provide resources for students to use to gather information (e.g., nonfiction books, brochures, CD ROMs, bookmarked sites on the Internet).

Review the research outline, and provide support as students look for relevant resource materials and conduct their research.

Once students have completed the research, have them display their findings in an interesting way, such as on a poster or in a book, and present their findings to the class.

Display the research projects in the classroom or the school library.

Activity Sheet

Directions to students:

Use the research outline to guide your research on your topic (1.10.1).

Extension

Read *The Lorax* by Dr. Suess. The story illustrates the importance of sustainable development and how humans must manage their use of plants to ensure survival.

Assessment Suggestions

- During class presentations, assess the students' research, as well as their oral presentation. Identify the main criteria for the research project, and list these criteria on the rubric on page 19. Use the rubric to assess the students' projects.
- As a class, discuss the criteria of good oral presentations. Use these criteria as the basis for students' self-assessment. You can have the students design a checklist for this self-assessment activity.

Date: ____________ Name: ____________________

Research Outline

Plant Product ____________________

Questions to research:

1. Which plant does the product come from?
2. Where is the plant grown?
3. How is the plant harvested?
4. What part of the plant is used in this product?

Also include:

1. a diagram of the plant
2. a diagram of the plant product
3. a list of resources used in your research

11 Making Dye From Plants

Materials

- assortment of plants or parts of plants (e.g., berries, beets, red and green cabbage, parsley, olives, tree leaves, flowers, beans, tree bark, orange, lemon, and tangerine peels, carrots, coffee, purple grapes)

Note: You may wish to have the students bring in one or two plants, or parts of plants, from the above list. This should provide you with a wide assortment of items and cut down on the cost of the experiment.

- an old, white sheet (cut into squares, approximately 15 cm x 15 cm)
- water
- pans for boiling water
- potato masher, fork, and sharp knife
- measuring cups
- heat source (adult supervision required)
- containers for dyeing cloth
- Popsicle sticks
- scissors
- glue
- newspaper
- smocks or paint shirts

Activity

Have students gather in a circle. Have them look at their clothing and the clothing of those around them, paying special attention to the colours in the clothing.

Ask the students:

- How do you think fabric for clothing is made into different colours?
- Have you ever slid on the grass and got a grass stain on your knee?
- Is a grass stain easy to wash off? Why or why not?
- Where do you think dyes come from?

Note: This activity may be done as a whole-class experiment, or at an activity centre with an adult supervisor.

Have students put on smocks or paint shirts and select a plant to use for dyeing. Have the students print the name of the plant on their activity sheet.

If possible, have the students mash their chosen plant (e.g., lemon peel) into a measuring cup. If a student is unable to mash the plant, have an adult cut it up, or leave it whole.

Measure 125 ml of the mashed plant and boil it in 500 ml of water for about five minutes, or until the water is coloured. Some plants may take longer, while others may require more water.

When the water is coloured, remove the pan from the heat source and let it cool. Remove the plant parts from the coloured water. The dye is now ready to use.

Place a square of the white fabric in each container. Pour dye in a container. Use a Popsicle stick to make sure all pieces of cloth are covered with the water.

After ten minutes, remove the fabric and cut a small piece from it. Let the piece dry on newspaper and then glue it onto the activity sheet.

Place the remaining fabric back into the container of dye, and leave the fabric in the containers overnight. Have the students remove the fabric the next morning (using the Popsicle stick) and lay it flat on the newspaper. Cut a small piece from the fabric, let it dry, and glue it onto the activity sheet.

Once all plants have been used to dye the fabric, compare how various plants produce colour on the fabric.

▶

11

Activity Sheet

Directions to student:

Write the name of the plant in the first column. In the second column, glue a small piece of the fabric that has soaked in the water for ten minutes. Glue a small piece of fabric that has soaked in the water overnight in the third column (1.11.1).

Extensions

- Have students mix several plants together to boil for dye. See how many different shades they can create.
- Dye pieces of wool or yarn and use them for future art projects such as weaving. Also trying dyeing fabrics to make a class quilt.

Date: ____________ Name: ____________

Making Dye From Plants

Plant	Colour of Cloth (after 10 minutes)	Colour of Cloth (overnight)

12 Plants and Animals Working Together

Materials

- art supplies, such as pencil crayons, felt markers, oil pastels

Activity

Divide the class into two groups. Designate one group Plants and the other group Animals. Have the groups stand on opposite sides of the classroom. Explain to the students that the goal of the game is for all of the students to join in the centre of the room. However, in order to move from one side of the room to the centre they must fulfill one requirement: each "animal" must say how he or she can help a plant, and each "plant" must say how he or she can help an animal. Encourage students to use first person (see examples below). Have the students form a circle when they come into the centre of the room.

Examples:

Plant: I provide food for animals. Food gives animals energy.

Animal: As a squirrel, I carry seeds for you from one place to another.

Plant: I am a tree, and I provide shade for you on a hot summer day.

Animal: I am a bee, and I help pollinate your flowers so that apples can grow.

Plant: You need me to make interesting colour dyes for your clothing.

Animal: I am a gardener, and I take care of plants.

If, toward the end, students are having difficulties coming up with responses, encourage them to seek help from others in their group.

Once everyone is in the centre of the room in a circle, ask the students:

- Why do you think we made a circle when we got to the centre of the room?
- How do plants and animals depend on each other?
- What would happen to plants if there were no animals?
- What would happen to animals if there were no plants?

Activity Sheet

Directions to students:

Illustrate one example of how animals depend on plants, and one example of how plants depend on animals. Be sure to explain your illustration in the space provided (1.12.1).

Extensions

- To reinforce plant and animal interdependence, create word cycles with the students. For example:

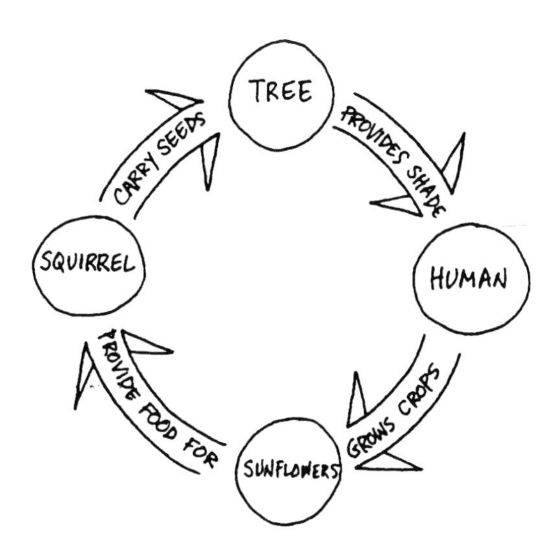

12

- Visit an apple orchard and discuss the importance of bees to the life cycle of an apple tree.
- Discuss the concept of *photosynthesis*. Research and discuss the importance of plants in producing oxygen for animals.

Activity Centre

Provide art materials for students. Have each student create one page for a big book called, *We Depend On Each Other.* Students may illustrate a picture of how an animal helps a plant, or how a plant helps an animal. Laminate the pages and bind the book together. Display the book in the classroom or in the school library.

If possible, have the students share the book with a grade-one class that is studying "Characteristics and Needs of Living Things."

Date: ____________________ Name: ________________________________

Plants and Animals Helping Each Other

How Plants Help Animals	How Animals Help Plants

13 Plants and Soil Erosion

Materials

- chart paper, felt markers
- 2 aluminum baking pans
- soil and sod
- water
- metric ruler
- paper cup
- permanent marker
- water bucket
- 5-cm to 15-cm thick wood block (wedge)
- 12 toothpicks

Activity

Review, from the previous activity, the various human uses of plants and plant products. Explain to the students that in addition to being of use to humans, plants also help the environment in many ways. Ask the students:

- In what ways could plants help the environment?
- What function do trees serve in our local area?
- Do you know of any other function of plants in our local area?

Record responses on chart paper. Ask the students:

- Have you ever built a castle on the beach?
- What happens to the castle when the waves hit it?
- Has anyone heard of the term *soil erosion*?

Explain to the students that the action of the water on the sand is similar to the way erosion works on the land. Explain that they are going to do an experiment to find out how soil erosion happens.

Using the activity sheet, introduce and work through the scientific method with the students. Discuss the purpose together, and have the students record the purpose; for example, "We want to find out how soil erosion happens."

Review the word *hypothesis*. Explain to students that a hypothesis is an "educated guess." Briefly explain what you are going to do during the experiment, and encourage students to hypothesize as to what they think will happen when water is poured over a container of soil and over a container of sod. The students can then record their hypothesis on the activity sheet.

Review the materials and the method together and record these on chart paper. Students can then refer to the chart paper when completing these sections of the activity sheet.

Fill the aluminum pan with packed soil until the soil is 2 cm deep. Now mark a line at the 2-cm point on the six toothpicks. Push the toothpicks into the soil in three rows (two toothpicks in each row). Tilt the pan and place the wood block underneath it to act as a wedge.

Take the paper cup and poke a hole in the bottom. When filled with water, the paper cup will simulate a cloud filled with rain. Sprinkle 500 ml of water over the soil. Use the bucket to catch the runoff. Repeat the process with the pan containing the sod.

Have the students record the results on their activity sheet. Ask:

- Which soil eroded the most?
- Why do you think this happened?
- How can plants reduce erosion?
- Where do you think most erosion occurs?
- Where could plants help control erosion?

13

As a class, discuss the conclusions of this experiment, and the applications to everyday life. Focus on the importance of plants in reducing erosion.

Activity Sheet

Directions to students:

Use the activity sheet to record your experiment on erosion (1.13.1).

Extensions

- Simulate erosion on different potted plants with various root systems. Compare the results of various plants.
- Invite a guest speaker from a local environmental organization to your classroom. Ask the guest to discuss environmental concerns such as erosion in your local area.

Date: ____________________ **Name:** __________________________________

Investigating Erosion

Purpose (what you want to find out): ____________________

____

____

Hypothesis (what you think will happen): ______________

____

____

Materials (what you will need): __________________

____

____

Method (what you did): __________________________

____

____

____

____

____

____

Results (what happened): ________________________

____

____

Conclusion (what you learned): __________________

____

____

Application (how you can use what you learned):

____

____

14 Protecting Plants

Materials

- chart paper, felt pens
- art paper (e.g., Manila tag)
- art supplies such as crayons, pastels, pencil crayons, glue
- samples from plants (seeds, twigs, leaves)
- pictures, magazines

Activity

Discuss what all living things need to stay alive. Brainstorm with the students. Ask:

- How are animals and plants the same? (Remind students that humans are animals.)
- What do all living things need to stay alive?
- Where do we get our water?
- Where do we get our food?
- Where does a plant get its water and food?

Focus on the importance of plants to humans. Ask:

- Do you remember the different ways plants help humans?
- How do plants help other animals?
- Why are plants important to humans?
- How are some plants destroyed? (Some examples are weather disasters, forest fires, excessive logging.)
- What can you do to replenish the plant supply in areas where damage has occurred?
- What would happen if you did`not plant new plants?
- What would happen if there were no plants left on Earth?
- How do humans make sure that plants are well looked after?
- Who takes care of all the different kinds of plants on Earth?

Discuss the various jobs related to plants, such as farmers, tree planters, flower growers, florists, nursery workers, and gardeners.

Have students create a poster showing ways in which humans can protect plants in order to preserve the environment. Display the posters throughout the school. You may wish to display them in April in celebration of Earth Day.

Extensions

- Read the book *Jen and the Great One,* by Peter Eyvindson. In the book, a tree tells the story of different peoples' interest in the forest.
- Have students interview those in your local community who have hobbies or careers related to plants. Encourage the students to share the results of their interview with the class.
- Take a field trip to a local wildlife centre or conservation area.
- Have students write letters to local government agencies about the importance of preserving and funding wildlife and conservation programs in your local community.

Assessment Suggestion

Have the students complete the student self-assessment sheet on page 22 to determine their knowledge and own actions for caring for plants.

References for Teachers

Bales, Suzanne Frutig. *Ready, Set, Grow!* Toronto: Macmillan, 1996.

Carratello, John, and Patty Carratello. *Hands On Science: Plants/Workbook*. Huntington Beach: Teacher Created Materials, 1994.

Conway, Lorraine. *Plants*. Superific Science Series. Carthage: Good Apple, 1980.

Davis, Beth. *Flowering Plants*. Grand Rapids: Instructional Fair, 1999.

Moore, Joellen. *Plants*. Monterey: Evan-Moor Corporation, 1998.

Westley, Joan. *Seeds and Weeds*. Sunnyvale: Creative Publications, 1988.

Unit 2

Magnetic and Charged Materials

Books for Children

Cole, Joanna. *The Magic Schoolbus and the Electric Field Trip*. New York: Scholastic, 1997.

Ehrlich, Robert. *Why Toast Lands Jelly-Side Down: Zen and the Art of Physics Demonstrations*. Princeton, NJ: Princeton University Press, 1997.

Gilman, Phoebe. *The Balloon Tree*. Toronto: Firefly Books, 1997.

Haas, Dorothy. *The Secret Life of Dilly McBean*. New York: Bradbury, 1986.

Lionni, leo. *Geraldine, The Music Mouse*. New York: Pantheon, 1979.

Locker, Thomas. *Anna and the Bagpiper*. New York: Philomel, 1994.

Meyrick, Kathryn. *The Lost Music: Gustav Mole's War on Noise*. New York: Child's Play, 1991.

Sandford, John. *The Gravity Company.* Nashville: Abingdon, 1998.

Simon, Seymour. *Einstein Anderson Shocks His Friends*. New York: Viking, 1980.

Van Allsburg, Chris. *The Wreck of the Zephyr*. Boston: Houghton Mifflin, 1983.

Waboose, Jan Bourdeau. *Morning On the Lake*. Toronto: Kids Can Press, 1998.

Web Sites

- **http://www.optonline.com/comptons/ceo/02936_A.html**

  Compton's On-Line Encyclopedia: extensive resource for researching magnets and magnetism. Includes informative articles (with many links) to magnetic poles, magnetic fields, magnetism and electricity, and the uses of magnets. With further references for magnets and magnetism.

- **http://library.advanced.org/11924/index.html**

  Think Quest's "The Wizard's Lab": this site includes basic information on electricity and magnetism, and answers questions, such as "what is electricity?" Shows the links between electricity and magnetism, and their applications (electric motors, CAT scans).

- **http://thunder.msfc.nasa.gov/**

  Lightning and atmospheric electricity research at Global Hydrology and Climate Center, working together with NASA: an introduction to lightning and how it affects space crafts, lightning properties, characteristics of a storm, types of lightning, and more.

- **http://www.mos.org/sln/toe/toe.html**

  Museum of Science, Boston: Theatre of Electricity: click on "Teacher Resources" for background information on static electricity (includes definitions of scientific terms), and activities to explore static electricity.

- **http://www.cleco.com/jfk_main.html**

  Electricity resource for students: excellent site for students researching electricity; includes safety tips, experiments, making electricity, and the history of the discoveries of electricity.

- **http://www.fi.edu/tfiexhibits/franklin.html**

  Permanent Ben Franklin Exhibit at the Franklin Institute: good site for students doing research on Benjamin Franklin. Includes information on his important contributions to science, especially in the area of electricity. Extensive links.

Introduction

In this unit, students will focus on materials that are magnetic and those that can hold an electric charge. Students will investigate the ways in which different materials affect magnetic strength and electric charge. They will learn that every magnet has two poles, and that the strength of a magnet depends on the types and combinations of the various materials from which it is made. Students will also describe their observations of static electricity and the conditions that affect it. Through these investigations, students will increase their knowledge about the properties of materials that make them useful for specific purposes.

Note: Investigations with static electricity work best when the air is dry. On humid days, charge on a conductor is reduced by the increased moisture in the air. Classrooms that have electric baseboard heaters, rather than forced air heat, tend to be less humid. If you have access to a hygrometer (humidity metre), readings under 30% lead to more successful experimentation.

General Information for Teachers

An example of an electrostatic force is when your hair stands on end after pulling a sweater over your head. A magnet picking up nails is an example of magnetic force. The magnet gets its name from Magnesia, in Greece, where the naturally magnetic mineral magnetite was discovered.

Note: Proper storage of the magnets is very important. If they are stored incorrectly, they could lose their charge. Ensure that the opposite poles are placed together. Also ensure that magnets are stored away from computers, computer disks, and computer tapes. Magnets can cause damage to these items.

Science Vocabulary

Throughout this unit, teachers should use, and encourage students to use, vocabulary such as: *north pole, south pole, attract, repel, charged, force, magnetic field, static charge, humidity, static electricity, magnetic poles,* and *geographical poles.*

Materials Required for the Unit

Classroom: magnets (bar and horseshoe), pencils, pens, erasers, paper clips, books (for weights), rulers, scissors, tape, string, chart paper, markers, large graph paper, paper, cardboard strips (0.5 cm thick), maze (included), centimetre graph paper (included), sink and faucet, Plasticine, water

Household: coins, buttons, combs, nails, tacks, screws, objects containing magnets (e.g., fridge magnets, electric can opener), metal needles, thread, towels, cardboard boxes, cork, wool mittens, tissue paper, aluminum foil

Equipment: hygrometer, tape cassette, audiotape (with recorded music on it), computer, computer disk (with file on it)

Other: copper wire, wax pieces, sawdust, Styrofoam chips, balloons, kettle or vaporizer or humidifier, silk cloth, wool cloth, glass rods, sealing wax rods, coat-hanger wire, blocks of wood, plastic wrap, rocks, small steel ball bearings, water table or basin, cardboard boxes, Plexiglas

1 Magnetic Attraction

Science Background Information for Teachers

Some metals are attracted to magnetic pull or force. Objects made from iron, cobalt, nickel, and some steel are attracted by magnets, while objects made from aluminum, copper, and tin are not affected. Objects made of other materials such as wood, plastic, and glass are also not attracted to magnets.

Materials

- a variety of magnets
- several objects, such as coins (pennies, nickels, dimes, and quarters), buttons, pencils, pens, combs, erasers, paper clips, nails, tacks, rocks, rulers

Activity

Divide the class into working groups and provide each group with various magnets. Give the students time to examine and discuss the magnets. Ask:

- What are these objects?
- What do they do?
- What kind of objects do they attract?
- How are magnets used in everyday life?

Provide students with several objects to examine. Tell the students they will be using these objects to test how they react to magnets; make sure the students do not start testing until you say to do so. Have the students discuss the characteristics of the objects, including the material each object is made of. On the activity sheet, have students record the name of each object and the material each is made of.

Now have the groups discuss how they think the objects will react to the magnets. Ask:

- Which objects do you think will be attracted to the magnets?
- Why?

Have the students record their predictions on the activity sheet and then test all items. Results can also be recorded on the activity sheet.

During follow-up discussion, ask:

- Were all the metal objects attracted to the magnets?
- Which objects were attracted to the magnets?
- Which objects were not attracted to the magnets?
- How are the objects that were attracted to the magnets similar?
- Why do you think some objects are attracted to the magnets?
- Could you pick up a nickel with the magnets? Why not? (Our nickel is made from an alloy – a combination of metals – so even though the metal nickel is attracted to magnets, there is not enough nickel in the coin for it to be picked up by the magnets.)
- If you wrapped some of the objects that were attracted to the magnets in cloth or other material, would they still be attracted to the magnets?
- Which part of the magnet attracts the best?

This discussion provides students with ideas of how magnets work, and which materials are attracted to magnets. The students will use their conclusions to complete the activity sheet.

1

Activity Sheet A

Directions to students:

Record the name of each object and the material the object is made of. Now predict which objects will be attracted to the magnet and which will not be attracted. Test each object and record the results (2.1.1).

Activity Sheet B

Directions to students:

Use the Venn diagram to show which materials are attracted to magnets (2.1.2).

Extensions

- Have the students take home a magnet and a copy of activity sheet A. Ask them to test twenty items in their home and record their predictions and results on the activity sheet. When they return the next day, have them share their findings with the class.
- Place items such as safety pins, paper clips, staples, and tacks in a glass of water. Have the students move a magnet up and down the outside of a glass. Do the objects move to the side of the glass and stick there as the magnet is moved up and down?

Activity Centre

Provide the following materials:

- index cards
- scissors
- coloured pencils
- tape
- variety of small objects (such as paper clips, buttons, bottle caps, and old jewellery) that can be tested for their reaction to magnets
- horseshoe magnet
- string
- pail or box
- metre sticks

Have the students draw fish on index cards, then cut out the fish. On the back of each fish, have the students tape on an object. The fish are now ready to be tossed in the pond (pail or box). Have students make fishing rods by tying one end of the string to a magnet and the other end to a metre stick. Tell them they are going to go fishing to test which "fish" are attracted to magnets and which are not attracted.

Have students lower their fishing rods into the "pool" and see which fish they can catch. They can record the results on activity sheet A.

Assessment Suggestion

Ask students to examine several objects to determine which are attracted to magnets and which are not (use objects different from those tested during the above activities). Have the students predict and sort the objects onto the Venn diagram (or print the name of each object on the Venn diagram). Ask students to focus on materials that are attracted by magnets, then explain their sorting rule to you. Use the anecdotal record sheet on page 15 to record results.

Date: ____________________ Name: ________________________________

Magnetic Attraction

Print the name of each object and the materials of which they are made. Use a ✓ to show your predictions and results as to whether or not each object was attracted to the magnet.

Object	Material	Prediction		Result	
		Attracted	Not Attracted	Attracted	Not Attracted

Date: ____________________ Name: ____________________

Materials and Magnets

attracted to magnets

not attracted to magnets

2 Magnets in Everyday Life

Science Background Information for Teachers

Magnets are used all around the house. You might find magnetic knife holders, and magnets to keep the fridge door and many cupboard doors closed. Magnets are used in all kinds of machines. They are in stereo speakers, television sets, computers, and telephones.

Magnets can be helpful if used properly. If used incorrectly, they can cause serious problems. Powerful magnets can actually erase the picture on a television screen by rearranging the atoms. They can also cause damage to cassette tapes and computer disks.

Recording tapes and computer disks are coated with an iron oxide. When we record music or sound on a cassette, the atoms on the tape are realigned so that when we play the tape back, the sounds will be reproduced. Similarly, the magnetic tape disk that turns around inside a floppy disk rearranges the atoms so that we can retrieve information at a later time. When a magnet is brought in contact with a cassette tape or computer disk, it mixes up the atoms so the sound or information cannot be retrieved.

Materials

- tape cassette with recorded music (Do not use a favourite tape as the music will be distorted. Prepare a tape just for this activity.)
- computer disk with a file of information on it (Do not use an important disk as the files will be erased. Prepare a disk just for this activity.)
- computer
- bar magnet
- tape player
- several objects containing magnets, such as fridge magnets, an electric can opener, and a wallet or purse
- chart paper, markers

Activity

Display the objects containing magnets, and have the students examine them. Then discuss the usefulness of magnets in everyday life. As a class, brainstorm uses for magnets and record these ideas on chart paper.

Demonstrate the harmful effects of magnets. First copy a file from the disk onto the computer. This will show students that there is a file on the disk. Now remove the disk from the computer and ask the students:

- What do you think will happen if I touch the bar magnet to the disk?

Take the bar magnet and pass it slowly over the computer floppy disk (about three to four seconds on the top of the disk, not over the metal wheel). Repeat this three or four times.

Now put the disk back in the computer, and try again to copy the file. You will likely receive a message telling you that you have an error or a statement saying that the computer cannot find the file. Ask the students:

- What do you think happened?
- What does this tell you about proper storage of computer disks and magnets?

Repeat this activity using the cassette tape. Play the music selection for the students. Ask:

- What do you think will happen if the magnet comes in contact with the tape?

Pass the magnet over the tape slowly at least five times.

Note: When you play back the tape in the tape player, parts of the tape will not be affected. Other sections will be completely blank, muffled, muddy sounding, or distorted.

Discuss the effects of the magnet on the tape, and the importance of proper storage of these materials.

Activity Sheet

Note: This activity sheet is to be completed at home under the supervision of parents/guardians. Remind students not to take household items apart, unless they receive permission and are supervised.

Directions to students:

Find as many objects as you can in your home that use magnets. Explain how the magnet is used in each item (2.2.1).

Activity Centre

Have students make their own fridge magnets. Provide several button-type magnets or magnetic strips and art supplies such as modelling clay, dried flowers, tree bark, shells, and beads.

Date: ____________________ Name: ________________________________

Magnets in My Home

List the objects in your home that use magnets.

Object	How the Magnet Is Used

3 Making a Magnet

Science Background Information for Teachers

The first magnets were actually pieces of rock. These rocks, called lodestones, contain magnetite. Magnetite is naturally attracted to iron. Magnetic induction occurs when a piece of metal like iron touches a magnet. It becomes a magnet itself. When a metal object containing iron or steel is rubbed by a magnet, the object will become magnetized. Iron will act like a magnet for a short period of time when treated like this. Steel (made from iron) will hold its magnetism for a long period of time.

Materials

- objects that are attracted to magnets (see Magnetic Attraction, page 85)
- bar magnets
- nails
- screws
- metal sewing needles
- paper clips (or staples)

Activity

Note: Have the students complete the activity sheet during the exercise.

Divide the class into working groups and provide all materials. Have the students examine a nail, and the paper clips or staples. Ask:

- Do you think the nail could pick the paper clips up?

Now have the students rub the nail fifteen times (in the same direction, not back and forth) with a bar magnet. Ask:

- Do you think the nail will pick up the paper clips now?

Have them test their predictions by testing the nail as a magnet. Have them record their predictions on the activity sheet. Ask:

- How many paper clips were attracted?

Have them record results on the activity sheet.

Now magnetize the metal sewing needle, then the screw, and repeat the exercise. Ask:

- How many paper clips were attracted to each of these?
- Which object worked best as a magnet?

Have the students record their predictions and results on the activity sheet.

Ask:

- Would it make a difference if the object was rubbed twice as much (thirty times)?

Have the students rub the objects thirty times on the magnet, then record their predictions and results on the activity sheet.

Activity Sheet

Directions to students:

Record the number of paper clips that the objects picked up after being rubbed with the magnet, then answer the questions (2.3.1).

Activity Centre

At the centre, provide paper clips (or staples), bar magnets, and objects such as nails, pens, pencils, and scissors to test. Have the students make a temporary magnet by touching the bar magnet to the head of the nail. Instruct them to use the nail's point to try to pick up the paper clips. Ask:

- How many paper clips can you pick up?

Now move the nail gradually away from the magnet. Ask:

- What happens to the paper clips?

Experiment with other objects such as pens, scissors, screws, and pencils.

Date: ____________ Name: ____________________

Making a Magnet

Record the number of paper clips the object picked up after being rubbed with a magnet.

Object	Number of Rubs	Number of Paper Clips Attracted	
		Prediction	Result

How can you tell when an object is charged or magnetized? ____________________

____

When you rub an object with a magnet, does the number of rubs make a difference to the charge?

____

____

4 Magnetic Force

Materials

- large graph paper
- markers
- horseshoe magnets
- bar magnets
- paper clips
- small steel ball bearings
- sheets of paper
- cardboard strips (5 or 6 per group) (The strips should be 0.5 cm thick. The length and width are not important.)
- centimetre graph paper (included) (Copy two sheets for each student or group.) (2.4.2)
- rulers

Activity A

Divide the class into working groups and provide each group with all materials. Have the groups pick up a paper clip with the bar magnet. Now challenge them to make a chain by continuing to pick up paper clips end to end. Have the students predict the number of paper clips that the bar magnet will hold in a chain. Continue to add paper clips until no more can be added without falling off the chain. Have the groups compare their results.

Repeat this activity using the small steel ball bearings instead of paper clips. Ask:

- Do you think the magnet will hold more or less ball bearings than paper clips?
- How many ball bearings do you think can be held in a chain?

Note: The magnetic pull will travel through both the paper clips and the ball bearings. At a certain point, however, the magnetism will be lost and the paper clips and ball bearings will no longer hold on.

As a class, create a bar graph to show the results of how many paper clips and ball bearings could be attracted to the magnet.

Activity B

Have the students work again in their cooperative groups. Each group should have five or six pieces of cardboard strips, a paper clip, a ruler, and the two different kinds of magnets (bar and horseshoe). Have the students place the bar magnet on the table. Have them then place the ruler on the table with the zero marking almost touching one end of the magnet. Have the students slide the paper clip along the edge of the ruler (without touching it), and bring the paper clip close to the magnet until a pull is felt.

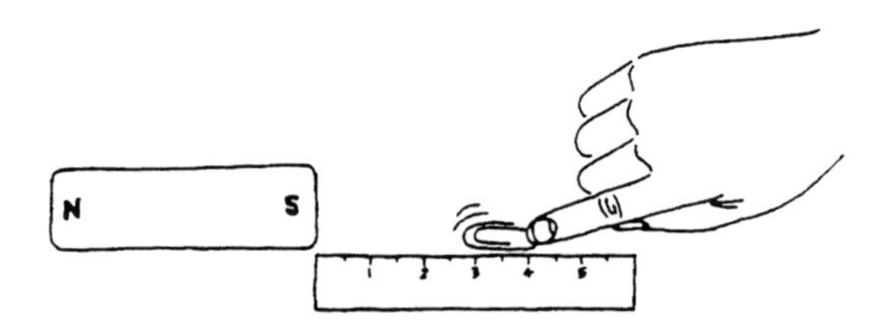

Have the students measure the distance between the paper clip and the magnet and record this on the activity sheet.

Now have the students examine one of the cardboard strips. Ask:

- If you placed the strip flat against the end of the magnet, do you think the magnet would still attract a paper clip?
- What if you added two, three, four, or five pieces of cardboard to the end of the magnet? Do you think the magnet would still attract a paper clip?
- How many pieces of cardboard do you think you could place between the magnet and the paper clip before no attraction was observed?

Have the students test their prediction. Repeat the activity but add a strip of cardboard between the magnet's end and the paper clip. Continue the tests for two strips, three strips, four strips, and five strips.

4

Students can continue to record results on the activity sheet. Ask:

- Does the magnet attract the paper clip through the cardboard?
- Does the attraction strength change when the amount of cardboard is increased?

Now repeat the investigation using the horseshoe magnet. Have the students test, measure, and record the results on the activity sheet.

Compare the bar and horseshoe magnets and have the students complete two bar graphs with their results, using the graph paper provided.

Note: Different magnets vary in strength. Therefore, results will vary when different magnets are tested.

Activity Sheet

Directions to students:

Measure and record the distance from the bar magnet to the paper clip when you observed an attraction/pull. Then measure and record the distance from the horseshoe magnet to the paper clip when you observed an attraction/pull (2.4.1).

Make two bar graphs to show your results for this investigation.

Note: Depending on your students' experience in constructing bar graphs, you may choose to guide this activity to ensure that students have included all components of a bar graph (see pages 9-10).

Extension

- Investigate other materials (lids from tin cans (edges taped for safety), plastic lids, aluminum pie plates, paper plates, staples, piece of cardboard) to determine what happens when they are placed between a magnet and an attracted object. Have the students hold a horseshoe magnet over some staples. Have them count how many staples were picked up. Now have the students remove the staples and spread them out again on the table top. The students can now test several flat items to see if the staples can still be picked up when other items are placed between the magnet and the staples.

Note: Surprisingly, items like the paper plate will have little effect (depending on the thickness), while the tin can (which contains iron) will actually cause the magnet to lose some of its magnetic ability. The reason is that the lid diverts the magnetic field, acting like a shield.

- Have the students test several other items to see if a magnet can move an attracted object through the item. Put a paper clip on a piece of cardboard. Hold the magnet under the cardboard and move the magnet around. What happens? Now put the paper clip on other materials such as heavy card paper, a metal 16 mm film container, a magazine, a book, a table top, a microscope glass slide, and a computer disk (empty, as the magnet will erase it otherwise).

Activity Centre

Use the following materials to make a magnet tester. (You may choose to make this prior to placing it at the centre, or you could have students each make their own as a part of the centre activity):

- 1 piece of wood, 15 cm long, 10 cm wide, and 2.5 cm deep
- 2 pieces of wood, 7 cm long, 10 cm wide, and 2.5 cm deep
- nails, hammer
- small magnet
- paper clip
- thread
- tape

▶

- materials such as paper, wax paper, aluminum foil, metal, plaster, overhead transparancy sheets, and glass microscopes

Nail the two smaller pieces of wood to each end of the longer piece to form the letter C. Tape the magnet to the underside of the top piece of wood. Tie one end of a 16-cm piece of thread to a paper clip. Tape the other end of thread to the base of the tester. Hold the paper clip upright under the magnet (there should be a gap of 1 or 2 cm between the magnet and the paper clip). It should now be suspended and appear to be free floating.

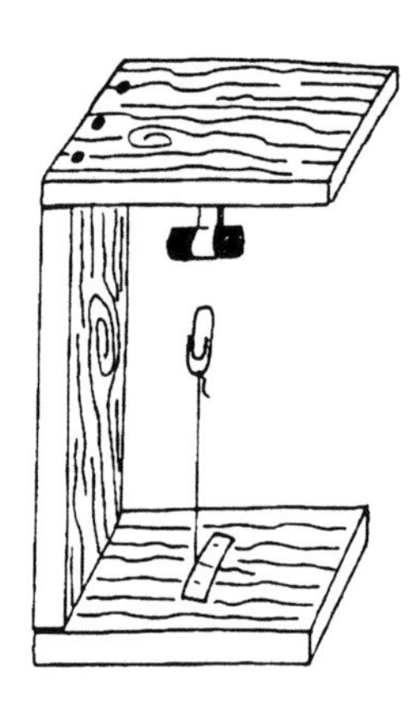

Have students use a variety of materials to place in the space between the paper clip and the magnet. These materials might include strips of paper, wax paper, aluminum foil, metal, plastic overhead transparency sheets, and glass microscope slides. Encourage students to test the effect of one piece of the material, as well as several layers of material. Have them predict and record on the activity centre sheet whether or not the material affects the magnetism (2.4.3).

Assessment Suggestion

Assess students' ability to construct the bar graphs. Identify criteria such as title, labelled x axis, labelled y axis, accurate data, and separated bars. Record these criteria and student results using the rubric on page 19.

Date: ____________________ Name: ______________________________

Magnetic Force

Measure and record the distance from the magnet to the paper clip when you observed a pull.

Number of Cardboard Pieces	Distance from Bar Magnet (cm)	Distance from Horseshoe Magnet (cm)
0		
1		
2		
3		
4		
5		

Use your results to make two bar graphs.

Date: ____________________ **Name:** ________________________________

Date: ____________________ Name: ________________________________

Floating Paper Clip

Material	Number of Layers	Effects - Paper Clip Floats (yes/no)	
		Prediction	Result

5 Designing and Constructing With Magnets

Materials

- paper clips
- sheets of paper
- water
- water table or basins
- towels (to dry up the water that may splash)
- bar magnets

Activity

Give each student several sheets of paper. Challenge them to take a piece and fold it into a boat that will float. Encourage students to try different designs. Allow them ample opportunities to discuss their designs with their classmates.

Have the students test their boats to see if they will float. Discuss and compare the boats to determine the factors that enable them to float. Ask:

- In which different ways could you make the boats move in the water?

Students may suggest wind power (a fan), blowing on them, or creating "waves" in the water to move the boat. Ask:

- Can you think of a way that you could use a magnet to move the boat?

Have each student select a boat that floats best, and attach paper clips to the "bow" and "stern" (front and back). Have them use a magnet to try to pull their boat across the water. Tell them they cannot touch the magnet to the paper clips, or aid the boat with their hands or by blowing on it. Upon completion of their test run, have the students design two other boats and test them out. Encourage them to investigate whether their boat moves better when more than one paper clip is placed at the bow and the stern.

Activity Sheet

Directions to students:

Draw a diagram of the boat design that moved best with the magnet. Label your diagram. Answer the questions to show what you have learned from this investigation (2.5.1).

Extensions

- Make boats out of balsa wood and place several nails or tacks in the base of the boats. Set a plastic serving tray of water on two piles of books or bricks. Have the students guide the boats by dragging a magnet under the tray.
- Read *The Wreck of the Zephyr,* a book by Chris Van Allsburg. Have the students write their own boat stories about their constructed boats.

Activity Centre A

Have the students design two boats: one made from Plasticine, and the other made from aluminum foil. Have the students place magnetic objects such as tacks, nails, paper clips, and ball bearings in the boats, then place the boats in a basin of water or in a water table. Have them hold a bar magnet in their hand under the water and try to guide the boat from one side to the other. While students are working, ask:

- Which boat worked best?
- Did the aluminum foil make a difference?
- Why?

Activity Centre B

Have students place some magnetic items (tacks, nails, paper clips) in an empty 500-ml or 1-litre soft drink bottle. Add some water to the bottle so that when the bottle is sealed and

5

placed in a basin of water, it settles under the surface of the water without sinking to the bottom. You have now created a "submarine." Using the magnet under the water, have the students guide the submarine from one side of the basin to the other without touching the bottle. Try various magnetic items inside the bottle to see which works best.

As a variation on this activity, have the students place a bar magnet inside the bottle, and use a straightened coat hanger to guide the submarine.

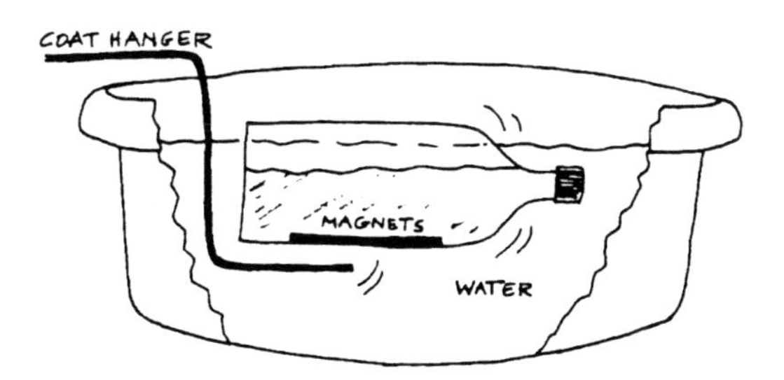

Date: ________________ Name: ____________________________

Building Boats

Draw a diagram of the boat design that moved the best with the magnet. Label your diagram.

Why do you think this design worked best? ________

____

____

If you did this activity again, how would you improve your boat design? ________________

____

____

6 Games With Magnets

Materials

- cardboard boxes with lids (shoeboxes work well)
- tape
- scissors
- ball bearings
- magnets
- cardboard
- paper sheets to cover the tops of the boxes
- maze (included) (2.6.1)

Activity

Begin the lesson by challenging the students to find their way through the maze, using their pencil to outline their route.

Now have the students work in cooperative groups to create a magnetic maze that a ball bearing will travel through. A box will form the foundation of the maze and ensure that the ball bearing stays inside. Students can design a maze on paper, then transfer the design to the base of the box. Have them cut out cardboard strips to use as walls for the maze, ensuring that there is a definite start and finish location. At the start and finish locations, have students cut a hole in the side of the box so that the ball bearing can be placed inside to start and will roll out at the finish.

Once they have completed their maze, have students place the ball bearing at the starting point. Have them test their design by dragging the magnet underneath the box to move the ball bearing through the maze.

Groups can now exchange their mazes and challenge one another. Have the students in each group report back on:

- what they learned
- what they would change in their model
- what things worked best
- what was the hardest part of the construction

Students can add to the challenge by putting the lids on their mazes and attempting to move the ball bearing through the maze with no visual assistance.

Activity Sheet

Note: This activity sheet is to be completed as an introduction to the above activity.

Directions to students:

Follow the path through the maze from start to finish (2.6.1).

Extension

Have students make scenic magnetic dioramas. You will need:

- shoeboxes
- file cards
- tape
- scissors
- paper clips
- magnets
- paint
- paintbrushes

Start by painting a scene (e.g., forest, desert, or beach) on the inside of a shoebox. Next have students decide what other things would be found in the habitat they have created (e.g., birds, fish, humans, clouds, sun, moon, mammals, insects). Have the students draw one or more of these figures on file cards, then cut them out. Tape a paper clip on the back of each figure. Now hold the cutouts against the back scene. Put a magnet on the back of the scene and let go of the cutout figure. The magnet should move the figure around the diorama.

It is now "showtime." Have students write and act out plays using their magnetic dioramas.

▶

6

Activity Centre

Fill a basin with sand or uncooked rice. Place at least twenty magnetic items in the basin (nails, paper clips, screws, steel teaspoons, steel thimbles, and so on) and mix them in the sand or rice so they are hidden. Challenge students to find the "buried treasures," using only magnets.

Assessment Suggestions

- Use the cooperative skills sheet on page 21 to record students' skills in working together.
- Have the students complete the cooperative skills self-assessment sheet on page 23.

Date: ____________________ Name: ______________________________

Find Your Way Through the Maze

7 Magnetic Fields and Polarity

Science Background Information for Teachers

On magnets, north and south poles attract each other. When you bring two north poles together they push each other apart. Similarly, two south poles will repel each other. This is one of the basic laws of magnetism. Opposite poles attract each other, while like poles repel each other.

Materials

- bar magnets (two for each group)
- string
- scissors

Activity

Review previous activities by asking:

- What effect do magnets have on staples and paper clips?
- Do magnets affect objects made from paper or plastic?
- How do you think magnets would affect each other?
- Would you expect two magnets to push on each other, pull each other, or do nothing?

Distribute the bar magnets and have students examine the N and S labels. Ask:

- What do you think these labels mean?

Have the students work in cooperative groups. Tie a string around the middle of a bar magnet. Ask one of the students in each group to hold the magnet by the string over a table. Have another student bring a second magnet close to the first magnet. Ask:

- What happens?
- Why do you think this happens?

Now have the students do the experiment according to the diagrams on the activity sheet, and complete the sheet.

When all groups are finished, have the groups report back their findings. Ask:

- Was there any evidence of a pulling force?
- Which parts of the magnet repelled (pushed away) each other?
- Which parts of the magnet attracted (pulled toward) each other?

Now have the students take the bar magnet with the string wrapped around the centre, hold it over the table again, and turn the magnet once in a clockwise or counterclockwise direction. Let the magnet rotate now on its own. When the magnet comes to rest, ask:

- Where is the magnet pointing?
- Which direction is this?

Have them try it two more times. Ask:

- Does the magnet point in the same direction each time?

Try this experiment in a different location in the room. Ask:

- Does the magnet still point in the same direction?
- What direction will the S of the magnet be pointing to?
- What direction will the N of the magnet be pointing to?
- How do the S and N markings on the magnets relate to directions on Earth?

Note: The north pole of the magnet will always point to Earth's North Pole.

Activity Sheet

Directions to students:

Set up your magnets as shown in the diagrams. Now draw diagrams to show what happens to the magnets (2.7.1).

Date: ____________ Name: ____________________

How Do Magnets React to Each Other?

Result

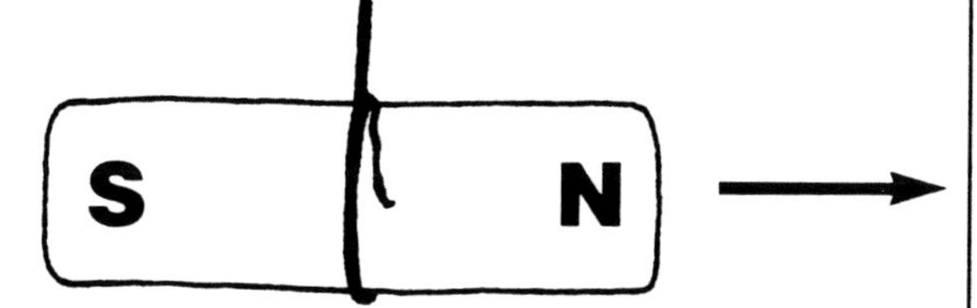

Result

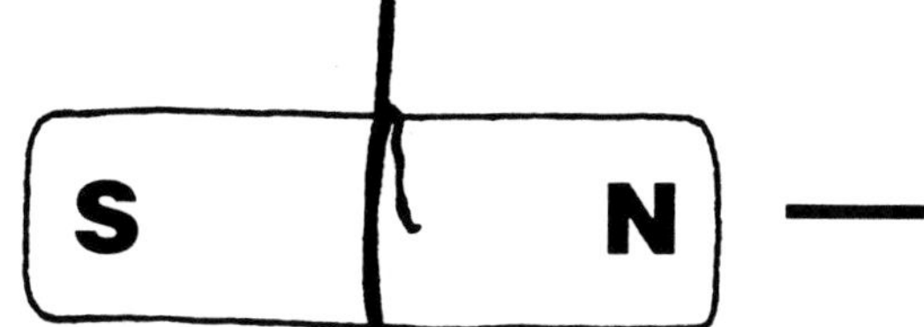

Result

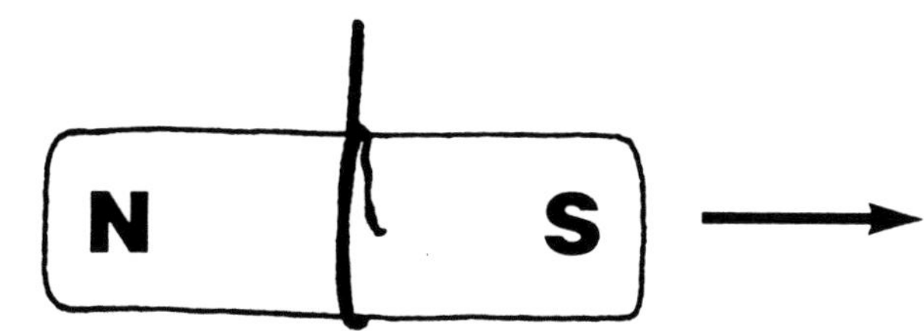

Result

Why do you think this happened? ____________________

____

____

____

8 Static Electricity

Science Background Information for Teachers

All matter is made up of atoms. Atoms are made up of smaller particles called *protons* (which carry positive charge), *electrons* (which carry negative charge), and *neutrons* (which carry no charge).

Static electricity is electricity that builds up in an object and stays there for some time. Static electricity can be produced when one object rubs against another. When you rub a balloon with a cloth, some electrons are transferred from the cloth to the balloon. The balloon becomes negatively charged. As in magnets, negative charges repel negative charges. If you take a negatively charged balloon and bring it in contact with an uncharged balloon, they attract each other. If you bring a charged balloon in contact with another charged balloon, they will repel (push away) each other.

It is easier to produce static electricity on cold, dry winter days. Moisture will pick up charges, making it more difficult to achieve good results during the following experiments.

Materials

- combs
- wool mittens
- tissue paper
- paper
- aluminum foil
- other materials such as copper wire, wax pieces, sawdust, thread, Styrofoam chips
- scissors

Activity

Have the students work in cooperative groups of three or four. In their groups, have them cut up the aluminum foil, tissue paper, and paper into small bits (about the size of a dime).

Have the groups spread the tissue paper bits out on another sheet of paper. Have one student from each group rub the comb briskly with the wool mitten, about ten times back and forth. (This should produce a static electricity charge.) Now have the students hold the charged comb near the tissue paper bits without touching the paper sheet, and observe what occurs. Ask:

- What happened to the tissue paper pieces?
- Why do you think this happened?
- What effect do you think rubbing the comb with the mitten had on the tissue paper?

Have the groups record their results on the activity sheet. Ask:

- What do you think will happen if you do the same test using paper bits and aluminum foil?

Have the students test their predictions by again rubbing the comb and observing its effects on the paper and on the aluminum foil. Results can be recorded on the activity sheet.

Have students prepare other materials to test. Cut thread into one-centimetre lengths, break up wax into pea-size pieces, strip copper wire and cut into one-centimetre pieces, and so on. All materials can now be tested to see how they react to the charged comb. Encourage students to make predictions before they test the materials, then record results on the activity sheet.

Following the investigation, ask:

- How did the different materials react to the comb?
- Which materials jumped onto the comb and stayed there?
- Which materials attached and fell and then reattached themselves?
- Did any materials try to attach themselves but could not be picked up?

▶

- Why do you think that happened?
- What materials did not react to the charged comb?
- How was the comb like a magnet?
- What do you think caused certain materials to be attracted to the comb?
- What caused the comb to attract materials?
- Do you think that rubbing the comb with the mitten did anything to the comb? What?

Discuss static electricity and introduce the word *charged* to explain that is what happened to the comb when it was rubbed with the mitten.

Activity Sheet

Directions to students:

During your investigations, use the chart to record your observations and the number of pieces of material attracted to the comb (2.8.1).

Extensions

- Have the students experiment with a charged balloon. Have them spread out small pieces of tissue paper across a table. Rub the balloon briskly for ten seconds with the wool mitten. Bring the balloon close to the tissue paper pieces to observe the reaction. Neutralize the balloon by wiping it with a damp cloth (make sure no water remains on the balloon). Now test the balloon by using materials other than the mitten, such as faux fur, a cotton or silk cloth, burlap, or hair. Students can record how much tissue paper was attracted each time, and create a bar graph of their results.
- Using the same materials as above, have the students try to attach the balloon to the wall, chalkboard, ceiling tile, and window glass. During investigations, encourage students to determine which surface and material worked best.

Assessment Suggestion

Have students complete the student self-assessment sheet on page 22 to reflect on their learning about static electricity.

Date: ____________________ Name: ________________________________

Static Electricity

Charging a Comb

Material Tested	Observations	Number of Pieces Attracted

Draw a diagram to show how you charged the comb.

9 Static Electricity and Humidity

Materials

- hygrometer (to measure humidity in the air)
- balloons
- permanent markers
- kettle, vaporizer, or humidifier
- water

Activity

Note: This activity will take place over the period of one school day, starting at the beginning of the day.

Show the hygrometer to the students. Explain that a hygrometer is an instrument that measures the amount (or percentage) of moisture or humidity in the air. Have the students read the hygrometer to determine the level of humidity in the classroom. Have them record this information on their activity sheet.

Give each student a balloon to blow up and fasten closed. Have them print their names on the balloons with the permanent markers. Ask:

- Is your balloon charged or uncharged right now?
- How can you charge this balloon?
- Do you think you can charge it so it is attracted to a surface like the wall?
- How long do you think it would stay charged and stick to the wall?

Have students use the activity sheet to record their predictions as to how long the balloon will stick to the wall.

Have students rub the balloon on their hair for ten seconds, then stick the balloon on the wall. Have them time how long it takes for the balloon to lose its charge and fall off the wall. (You can continue with other classroom activities while this test is being conducted.) Have students record this information on the activity sheet.

Note: If the humidity level of the room is very low, the balloons may stay on the wall for a day or longer. Usually the balloons fall down after fifteen or twenty minutes (depending on the size and weight of the balloon).

Now ask the students if they can think of ways to increase the humidity level in the classroom. Use a humidifier, vaporizer, or boiling water in a kettle to do this over a period of an hour. The classroom door should remain closed in order to maintain the moisture. Ask:

- Do you think that increasing the humidity will affect the charge of a balloon?
- If you rubbed the balloon again on your hair, would it stick to the wall for a longer or shorter period of time?

Have the students record their predictions on the activity sheet. After a few hours, once the humidity level in the room has increased, have students record the new humidity level on their activity sheet. Conduct the investigation again, having students measure the amount of time the balloons stick to the wall. Have them record the results. Ask:

- How did the humidity affect the charge of the balloon?

Students can now complete the activity sheet.

Activity Sheet

Directions to students:

Record the results of your investigation on your activity sheet (2.9.1).

Extensions

- Have students write up the above activity as an experiment, using the extension activity sheet (2.9.2). Depending on your students' background experience writing up activities using experimental design, you may choose to guide this activity.
- Read *The Balloon Tree*, a book by Phoebe Gillman.

Date: ____________ Name: ____________________

How Long Will a Balloon Stay on the Wall?

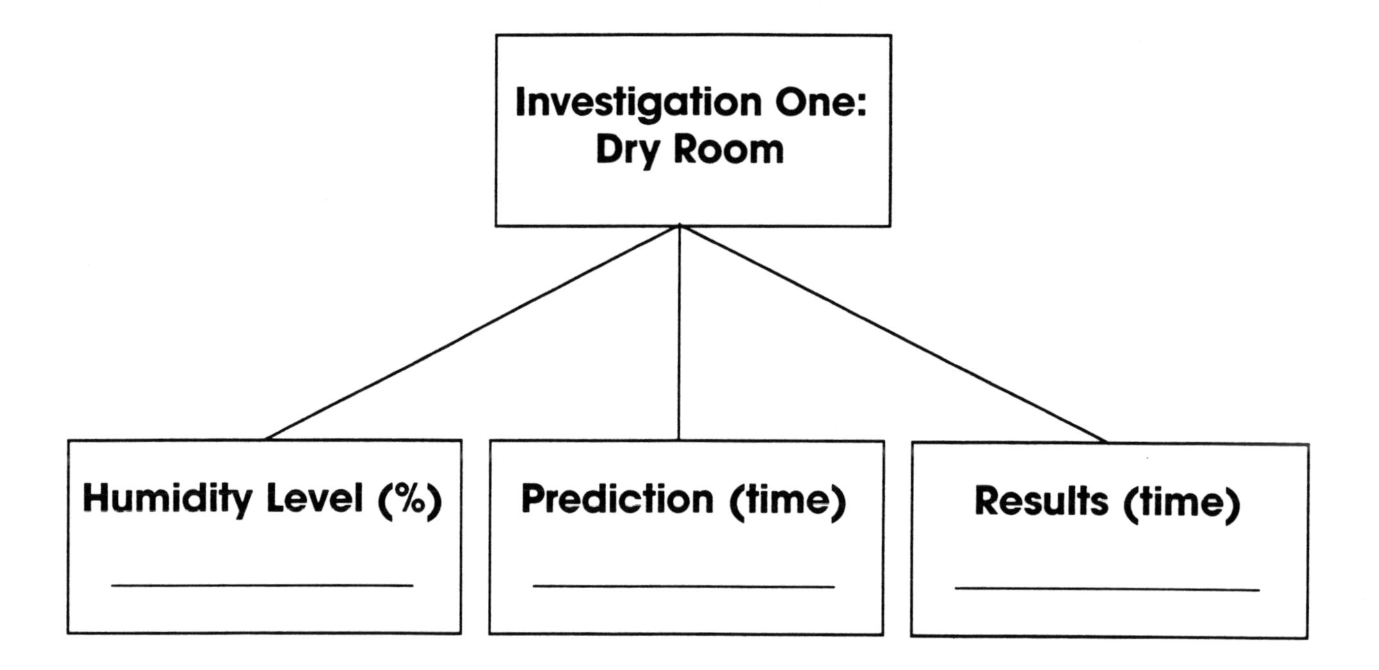

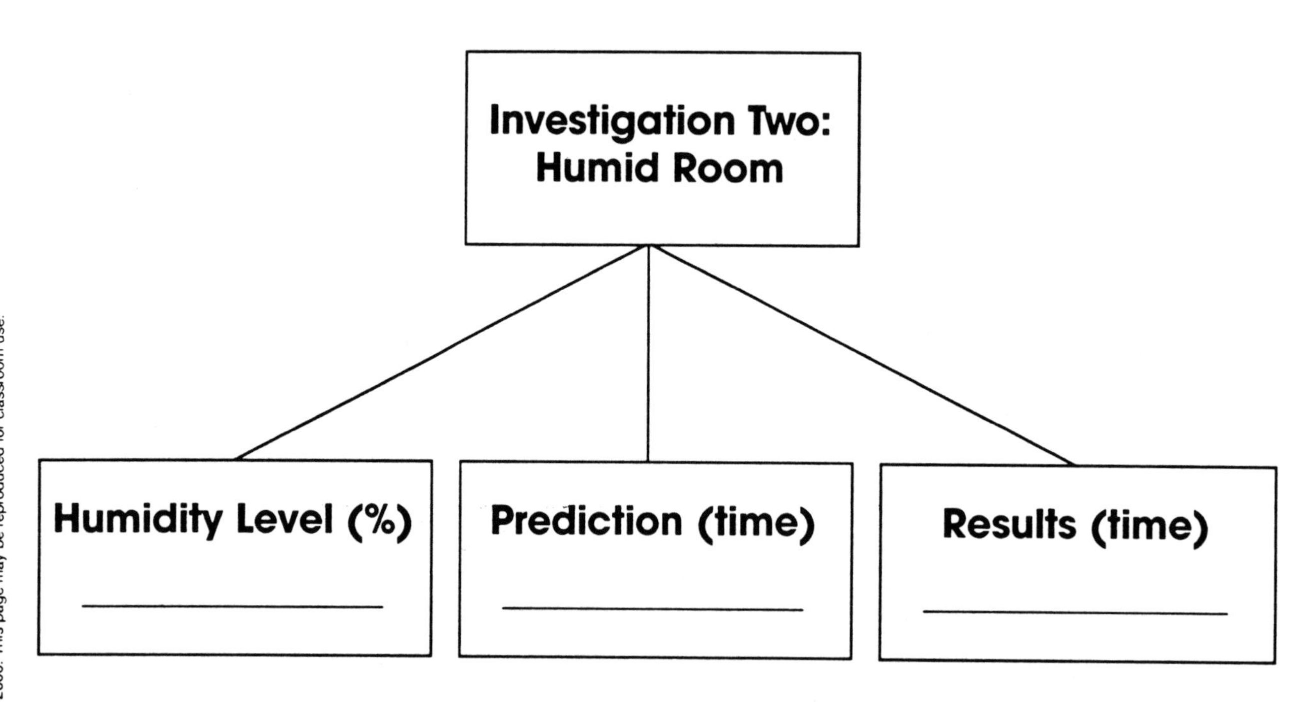

Date: ______________ Name: ______________________

Experimenting With Static Electricity and Humidity

Purpose (what you want to find out): ______________

____

____

Hypothesis (what you think will happen): __________

____

____

Materials (what you need): ____________________

____

____

Method (what you did): ________________________

____

____

____

____

____

Results (what happened): ______________________

____

____

Conclusion (what you learned): _________________

____

____

Application (how you can use what you learned):

____

____

10 The Force of Static Electricity

Science Background Information for Teachers

There are two kinds of static electrical charges, positive and negative. Like charges repel each other; unlike charges attract each other. Rubbing various objects with certain materials produces a charge. Even the flow of water can be attracted to charged objects.

Rubbing can also occur as air moves up through a cloud. When the water droplets in the cloud fall in the air, a charge can build up in the cloud. If unlike charges build up in different parts of the cloud, a bright spark (lightning) may jump across it. This will neutralize the cloud. Lightning will not occur again until the same conditions in the cloud are recreated.

Similarly, when you walk across a carpet your body becomes charged because your feet rub the carpet. A spark can fly between you and a metal light switch or doorknob if the metal is uncharged. Once this occurs, your body is discharged and you can touch that object again and not receive the shock.

Materials

- sink and faucet
- silk cloth
- wool cloth
- comb
- 2 glass rods
- string
- 2 sealing wax rods

Activity: Part One

Note: This activity works best as a demonstration since it is unlikely there will be enough sinks and taps for cooperative groups.

Turn on the water faucet so that a thin stream of water is flowing. Place the uncharged glass rod near, but not touching, the flow of water. Ask students:

- Is the rod affecting the water flow?

Now rub the glass rod with the silk cloth. Ask the students:

- Is the rod now charged or uncharged?
- What do you think will happen if the rod is brought close to the water stream?

Test their predictions by bringing the charged rod near, but not touching, the flowing water. Observe and discuss the results.

Note: The water stream will bend toward the rod because it is attracted to the charged object.

Repeat this investigation using a wool cloth and a comb, and compare results.

Activity: Part Two

Note: The following activity works well as a demonstration.

Tie a string so that it has a loop at each end. Charge the two glass rods by rubbing them with the silk cloth. Have a student hold the string at the centre and place one of the glass rods through the loops in the string, as shown in the diagram:

Now bring the other rod near one end of the suspended glass rod. While students observe carefully, ask:

- What happens? (Since the rods have the same charge, they repel each other.)

Repeat this using the wax rods instead of the glass rods. Finally, try one glass rod with one wax rod. Discuss and compare the results of this investigation, and have students complete the activity sheet.

Activity Sheet

Directions to students:

Draw diagrams of the two investigations (2.10.1).

Extensions

- Group students in pairs. Have each student blow up a balloon, then attach a string to the balloon. Have the students predict what will happen when they hold the strings and bring the two balloons together. Now have the students rub one balloon against their hair ten times, predict what will happen when the balloons are brought together, then test what does happen when the two balloons are brought together. Finally, have the students rub both balloons against their hair ten times each, predict, and test.
- Tell the students the story of St. Elmo's fire. St. Elmo's fire is named after St. Erasmus, the patron saint of sailors. A long time ago, sailors were sometimes frightened at night by a strange bluish light that danced on the masts of their ships. The light was often accompanied by crackling noises. Airplane pilots sometimes see St. Elmo's fire on the wings and propellers of their planes. It usually happens when they are flying near thunderstorms or cumulonimbus clouds. Static electricity often builds up on tall objects when the atmosphere is full of electrical charges. St. Elmo's fire is really a discharge of static electricity, and is usually seen or heard before or during a thunderstorm.
- Discuss the hydrogen-filled airship, the *Hindenburg,* which is believed to have exploded because of a discharge of static electricity near to a hydrogen gas leak.

Note: Students may be interested in doing further research on St. Elmo's fire and the *Hindenburg* to determine the role that static electricity played in each.

Date: ____________________ Name: ________________________________

Draw diagrams of your investigations, and explain each diagram.

Charged Rods and Flowing Water

Diagram	Explanation

String and Glass Rods

Diagram	Explanation

11 Making an Electroscope

Science Background Information for Teachers

An electroscope is a device used to detect and identify static electricity.

Materials

- aluminum foil
- corks (cork balls work better than wine corks, but both work)
- 30-cm pieces of coat hanger wire
- 15-cm pieces of thread
- Plasticine
- blocks of wood, 2.5 cm x 10 cm x 10 cm, with a predrilled hole in the centre of the block (Drill a small hole slightly larger than the hanger wire's diameter.)
- thumbtacks
- wax rods
- combs
- glass rods
- balloons
- silk cloths
- wool cloths
- plastic wrap

Activity

Electroscope Model

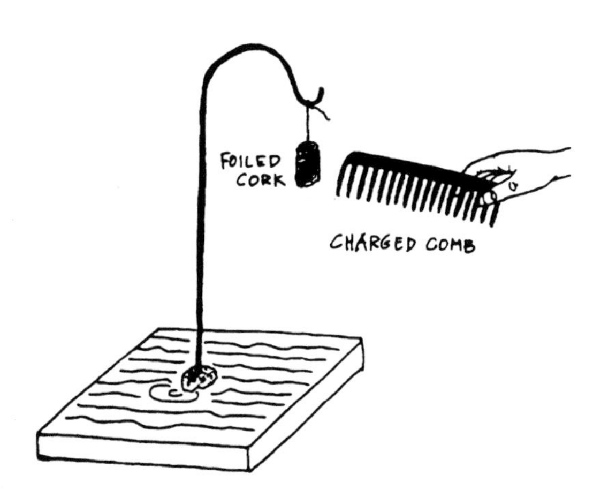

Making an electroscope:

1. Wrap aluminum foil around the cork.
2. Loop one end of the thread several times around the nail of the thumbtack. Press the thumbtack into one end of the cork wrapped in foil.
3. Bend a coat hanger to look like a street light. (You may need to use pliers to bend the top end into a hook.) Place the other end of the hanger into the hole in the block of wood. Secure with Plasticine.
4. Hang the cork from the hook and tie a knot in the thread to secure it to the hook.

The electroscope is now ready to be tested. To detect and identify static electricity, bring objects close to the foil-wrapped cork. As you bring a charged object near the cork, the cork will be attracted to it. If the cork and the object repel each other, this means that both are charged. You can eliminate the charge on the cork by touching it with your hand to ground it after every test. If there is no reaction between the object and the cork, neither is charged.

Have the students test several objects such as the wax rods, glass rods, balloons, and comb. Rub each object with a variety of materials such as the plastic wrap, wool, and silk. Students can record predictions and results on the activity sheet.

Activity Sheet

Directions to students:

Use the chart to record your predictions and results of the investigations with your electroscope (2.11.1).

Activity Centre

Have students use an electroscope to measure if distance from an object affects the attraction of a charged object. Have them conduct five trials, each time changing the number of times they rub a glass rod with a silk cloth (two, four, six, eight, and ten rubs). Have the students use their ruler to measure the distance a charged glass rod is from the cork. Have them start with the charged rod 20 centimetres from the cork, then slowly close the distance until the cork is attracted. With a ruler, they can measure the distance at which the cork is attracted to the rod. After each trial, use the damp cloth to neutralize the charge. Record the results on the activity centre sheet (2.11.2).

Assessment Suggestion

Have individual students demonstrate how the electroscope works. Through questioning, assess their ability to explain the concepts and use the related vocabulary (charge, attract, repel). Use the individual student observations sheet on page 16 to record results.

Date: ____________________ Name: ________________________________

Testing the Electroscope

List the names of the objects you are testing.
List the material used to rub each object.
Record your predictions and results using the terms *attracted, repelled,* or *nothing.*

Object	Material Rubbed With	Prediction	Result

Date: ____________ Name: ____________________

Rub the glass rod with a silk cloth the number of times shown on the chart. Measure the distance at which the glass rod is attracted to the cork.

Charging: Number of Rubs	Distance (cm)
2	
4	
6	
8	
10	

Draw a diagram of your electroscope. Label the parts.

12 Static Electricity in Everyday Life

Materials

- chart paper, markers
- 30-cm square sheet of Plexiglas
- 4 books
- Styrofoam packing chips
- wool cloth
- silk cloth
- plastic wrap
- centimetre graph paper (included) (2.12.2)

Activity

Now that students have had some experience with static electricity, have them discuss examples of static electricity that they have observed in everyday life. Record their examples on chart paper (and refer to them during the rest of the unit). These examples may include:

- hair standing on end after pulling a sweater over their head
- hair crackling and standing on end when it is combed
- feeling a "shock" after walking across a carpet and touching an object such as a lamp
- window decorations that cling to glass
- dusters and dust mops that attract dust particles
- clothes clinging together after being removed from a dryer

Discuss ways in which static electricity can be helpful. Ask:

- Think back to the activity you did when the sawdust was attracted to the charged comb (static electricity on page 108). How could you use this knowledge to help you in everyday life? (Some examples are: dust mops, blind and curtain cleaners.)

To have the students further understand this concept, divide them into working groups and provide each group with the materials listed. Have the students scatter several Styrofoam packing chips over the surface of their table. Ask:

- How could you pick these up without touching them?
- How could you use static electricity to do this?

Have the students rub the Plexiglas with the wool cloth, then turn the Plexiglas over so that the charged surface is facing down. Now, move the Plexiglas near the Styrofoam chips.

Note: The charged Plexiglas will attract the chips and pick them up.

Now have students experiment to determine which material charges the Plexiglas best in order to pick up the Styrofoam chips. The students can use the activity sheet to record their experiment and then create a graph of the results using the graph paper provided.

Activity Sheet

Directions to students:

Record the results of your experiment and make a graph of your results (2.12.1).

Date: ________________ Name: ____________________________

Experimenting With Charged Plexiglas

Purpose (what you want to find out): ______________

Hypothesis (what you think will happen): __________

Materials (what you need): ____________________

Method (what you did): ______________________

Results (what happened): ____________________

Conclusion (what you learned): ________________

Application (how you can use what you learned):

Date: ____________________ **Name:** ________________________________

References for Teachers

Bosak, Susan. *Science Is…* . Richmond Hill, ON: Scholastic, 1991.

Catherall, Ed. *Fun With Magnets*. Hove, UK: Wayland, 1985.

Challoner, Jack. *My First Batteries & Magnets Book*. New York: Dorling Kindersley, 1992.

Farndon, John. *How the Earth Works*. Pleasantville, NY: Readers Digest Association, 1992.

Levenson, Elaine. *Teaching Children About Physical Science*. New York: McGraw-Hill, 1994.

Strongin, Herb. *Science on a Shoestring*. Menlo Park, CA: Addison-Wesley, 1985.

Unit 3

Forces and Movement

Books for Children

Anderson, Margaret Jean. *Isaac Newton: The Greatest Scientist of All Time*. Springfield, NJ: Enslow Publishers, 1996.

Bedard, Michael. *The Nightingale*. New York: Clarion Books, 1991.

Cobb, Vicki. *Why Doesn't the Earth Fall Up?: And Other Not Such Dumb Questions About Motion*. New York: Lodestar Books, 1988.

Gardner, Robert. *Experiments With Motion*. Springfield, NJ: Enslow Publishers, 1995.

Haas, Dorothy. *The Secret Life of Dilly McBean.* New York: Bradbury Press, 1986.

Hewitt, Sally. *Forces Around Us*. New York: Children's Press, 1998.

Lafferty, Peter. *Force & Motion*. New York: Dorling Kindersley, 1992.

Lionni, Leo. *Alexander and the Wind-up Mouse*. New York: Pantheon, 1969.

Madgwick, Wendy. *Magnets & Sparks*. Austin, TX: Raintree Steck-Vaughn, 1999.

Murphy, Bryan. *Experiment With Movement*. Minneapolis: Lerner Publication, 1991.

Oxlade, Christopher, and Andrew Farmer. *Energy and Movement*. New York: Children's Press, 1998.

Parker, Steve. *Magnets*. Milwaukee: Gareth Stevens, 1998.

de Pinna, Simon. *Forces and Motion*. Austin, TX: Raintree Steck-Vaughn, 1998.

Riley, Peter D. *Forces and Movement*. New York: Franklin Watts, 1998.

Sandford, John. *The Gravity Company*. Nashville: Abingdon Press, 1988.

Skurzynski, Gloria. *Zero Gravity*. Toronto: Maxwell Macmillan, 1994.

Web Sites

- **http://www.optonline.com/comptons/ceo/02936_A.html**

  Compton's On-Line Encyclopedia: extensive resource for researching magnets and magnetism. Includes informative articles (with many links) to magnetic poles, magnetic fields, magnetism and electricity, and the uses of magnets. With further references for magnets and magnetism.

- **http://library.advanced.org/11924/index.html**

  Think Quest's "The Wizard's Lab": this site includes basic information on electricity and magnetism, and answers questions such as "what is electricity?" Shows the links between electricity and magnetism, and their applications (electric motors, CAT scans).

- **http://thunder.msfc.nasa.gov/**

  Lightning and atmospheric electricity research at Global Hydrology and Climate Center, working together with NASA: an introduction to lightning and how it affects space crafts, lightning properties, characteristics of a storm, types of lightning, and more.

- **http://www.mos.org/sln/toe/toe.html**

  Museum of Science, Boston: Theatre of Electricity: click on "Teacher Resources" for background information on static electricity (includes definitions of scientific terms), and activities to explore static electricity.

- **http://www.cleco.com/jfk_main.html**

  Electricity resource for students: excellent site for students researching electricity; includes safety tips, experiments, making electricity, and the history of the discoveries of electricity.

- **http://www.fi.edu/tfiexhibits/franklin.html**

  Permanent Ben Franklin Exhibit at the Franklin Institute: good site for students doing research on Benjamin Franklin. Includes information on his important contributions to science, especially in the area of electricity. Extensive links.

- **http://ekimo.com/~billb/emotor/sticky.html**

  Science Hobbyist: to help explain electricity, this site provides ten experiments using plastic tape. These interesting activities provide information about static, positive and negative attraction, and charges. Included is a reference section to help you learn more.

- **http://www.howstuffworks.com/index.htm**

  How Stuff Works: type "gravity" in the search box and find the answers to many questions on this or any topic. The information is accessible and the site provides many helpful links.

- **http://nyelabs.kcts.org/openNyeLabs.html**

  From the television show "Bill Nye the Science Guy": click on teacher's lounge and then episode guides to find information on many different topics. Lessons plans and links to the episodes are great for teachers and students.

- **http://www.treasure-troves.com/bios/Newton.html**

  Learn more about the father of physics, Sir Isaac Newton. This site has good information about his life, work, and many extensions and explanations.

Introduction

This unit introduces students to two types of forces and the effects of these forces. The first type of force involves pushes and pulls, or direct interaction. The second type involves interaction at a distance, such as magnetic or static electric forces. The students will explore the effects of different forces including muscular force, magnetism, static electricity, and gravitational force. Students will investigate ways in which forces create movement in objects and will expand their understanding by designing and making devices that use a form of energy to create controlled movement.

The study of the effects of magnetic and static electrical forces is introduced in the Grade Three Matter and Materials unit titled Magnetic and Charged Materials. Several of the activities in this unit on Forces and Movement deal with similar concepts. These activities can be used with students to reinforce ideas already introduced in the unit on Magnetic and Charged Materials.

The concept of forces can be a difficult topic for students this age to understand. When possible, provide hands-on examples of devices or pictures of devices that use various forms of energy to function and create movement (for example: windup toys, paper airplanes, remote-control toys, and small household appliances).

Science Vocabulary

Throughout this unit, teachers should use, and encourage students to use, vocabulary such as: *force, push, pull, direct force, muscular force, indirect force, load, distance, speed, magnetism, static electricity, gravity, function, friction*, and *movement*.

Materials Required for the Unit

Classroom: chart paper, felt markers, mural paper, pencils, magnets (different sizes and shapes), erasers, scissors, rulers or metre sticks, pencil crayons, dictionary, books, string, graph paper (included)

Books, Pictures, and Illustrations: pictures of everyday devices (from magazines and newspaper flyers)

Equipment: spring scale (that measures in newtons)

Other: keys, water bottles, a small bag with handles, cardboard box filled with light objects, cardboard box filled with heavy objects, knapsack filled with books, empty thread spools, beads, elastic bands, toy cars, balloons, pieces of cloth (wool, flannel, or felt), weights (or fairly heavy objects), ball, nuts, washers, variety of surfaces (e.g., smooth flooring, carpeting, sandpaper), variety of toys that store energy to create movement (e.g., jack-in-the-box, plane propelled by elastic band, windup toys), wooden block (with an eye screw in one side)

1 Forces – Push or Pull

Materials

- cardboard box filled with light objects
- cardboard box filled with heavy objects
- knapsack filled with books
- chart paper, felt markers
- spring scales measured in newtons
- various everyday objects (e.g., keys, water bottle, scissors)
- small bags with handles
- toy cars
- dictionary

Activity A

As an introduction to the lesson, provide each student with a toy car. Challenge students to move the toy car in as many ways as they can. Invite them to discuss the ways in which they were able to move the cars.

Now place the cardboard boxes and knapsack on the floor in the classroom. Have the students make a circle around the boxes and knapsack. Ask the students:

- If I want to move the knapsack from one side of the circle to the other without picking it up off the floor, how would I do it?

Select a student to move the knapsack. Ask:

- What direction did the knapsack move?
- Was the knapsack pushed or pulled?

Select another student to move the light cardboard box. Give a number of specific instructions as to how the box is to be moved (for example: push the box to the left, pull the box to the right, push the box slowly forward, pull the box quickly backward). Repeat with the other box. Discuss the difference in moving the box filled with light objects versus the box filled with heavy objects.

Ask the students:

- How did you move the knapsack and the boxes?
- What did you need in order to move the boxes and knapsack from one spot to another?
- What is a force?

Have the students brainstorm several sentences containing the word *force*. Record these on chart paper. This activity highlights the many meanings of this term. Students, for example, may create sentences such as, "May the force be with you" or "I will force you to do it." After students have made up several sentences, have them look up the term *force* in the dictionary. Record the definitions on chart paper.

Focus on the scientific definition. Explain to the students that a force is a push or a pull. Forces often produce movement. Things stay still unless a force pushes or pulls them.

Have students brainstorm other examples of force in the classroom. Record these answers on chart paper (e.g., pushing a door open, pulling a book from a shelf, pushing a pencil across a piece of paper, dragging or pulling a gym bag along the floor).

Activity B

Ask the students:

- How do you measure the length of an object?
- How do you measure the mass of an object?
- Do you think you can measure the force of an object?

Introduce the newton scale. Explain to the students that a unit of force was given the name *newton* in honour of Sir Isaac Newton, who discovered the nature of forces.

1

Divide the class into groups of three or four students. Allow them some initial time to experiment with the newton scale. Have one student sit in a chair. Have another student hook the scale onto the chair and hold the scale in place. Encourage the students to use the spring scale to get a feel for 1 newton, 5 newtons, 10 newtons, and 15 newtons of force by pulling on the hook of the scale.

Once the students have had an opportunity to experiment with the newton scale, have them use the activity sheet to record their investigations. Have them place a bag on the hook of the spring scale. (The bag should be resting on the ground.) Have the students place an object in the bag, record the name of the object on their activity sheet, estimate how many newtons will be necessary to lift the bag, and record their estimate on their activity sheet. Then have them lift the bag with the object and record the actual number of newtons required to lift the bag.

Now have the students place the bag on the ground and hook the scale on the handles. Have them guess how many newtons it will take to pull the bag along the ground. Make sure students record their estimates on their activity sheet. Once their estimates have been recorded, have the students pull the bag along the floor and record the actual number of newtons required.

Have the students repeat the activity using a variety of different objects inside the bag.

Activity Sheet

Note: This activity sheet is to be completed during the activity.

Directions to students:

Record the names of the objects measured, as well as your estimates and results (3.1.1).

Extensions

- Have the students repeat the above activity, but use different surfaces to pull the bag on (e.g., tile floor, carpet, grass, and sand). Have students determine whether or not the surface has an effect on the amount of force required.
- Have students research the life of Sir Isaac Newton.

Activity Centre

Encourage students to collect objects from home and from around the classroom. Provide spring scales and additional activity sheets. Encourage students to use the spring scale to measure the force in newtons to lift or pull the various objects collected.

Assessment Suggestion

As students are working with the spring scale, observe their ability to estimate and measure. Also ask students to define force and to provide examples of force using the materials. Use the anecdotal record sheet on page 15 to record results.

Date: ____________ Name: ____________________

Investigating Forces - Push and Pull

Object	Action	Estimate Force (N)	Actual Force (N)
	Lifting		
	Pulling		
	Lifting		
	Pulling		
	Lifting		
	Pulling		
	Lifting		
	Pulling		
	Lifting		
	Pulling		
	Lifting		
	Pulling		

2 Friction

Materials

- wooden blocks with an eye screw in one end
- weights or heavy objects
- spring scales that measures in newtons
- surfaces on which to test friction (e.g., smooth flooring, carpeting, and large pieces of sandpaper)

Activity

Begin the activity by having the students rub their hands together. Ask the students:

- What do you notice when you rub your hands together rapidly? (They should feel the warmth caused by friction.)
- Why do you think your hands started to feel warm?
- Does anyone have an idea what you have created by rubbing your two hands together?

Explain to the students that they have created friction. Friction is created when two surfaces rub together. Friction can be increased or reduced depending on the surfaces of the two objects that come in contact with each other.

Tell the students that they are going to conduct an experiment to find out the forces needed to pull a block across different kinds of surfaces (a smooth surface like tile or linoleum, a carpeted surface, and the sandpaper). Ask:

- What device did you use to measure forces during the previous activity?
- What unit is used to measure force?

Divide the class into working groups and review the procedure for the investigation.

1. Attach the spring scale to the eye screw.
2. Place the wooden block on the floor. Estimate the number of newtons required to pull the block across the smooth surface.
3. Pull the block gently at a constant speed over the smooth surface. Record the number of newtons required on the activity sheet.
4. Repeat with the carpeted surface and the sandpaper surface.
5. Once again, record the results.
6. Now place a weight on top of the wooden block and repeat the experiment using all three surfaces. Record the results.

Have students share their findings in a large group once they have completed the investigation. Ask:

- What type of surface increased the amount of friction?
- What type of surface decreased the amount of friction?
- Did your results vary when you placed the weight on the wooden block? If so, how?

Activity Sheet

Note: The activity sheet is to be completed during the activity.

Directions to students:

Record your results on the chart and answer the questions at the bottom of the page (3.2.1).

Extensions

- Have students design and construct a maze or obstacle course with a surface that will allow the wooden block to be pulled using the least or most newtons.
- Collect pictures from car magazines that show cars being driven on various road surfaces. Have students identify the safest and most dangerous road surfaces. Also discuss how tires help create the friction necessary to drive and stop cars on slippery surfaces.

2

Assessment Suggestions

- As students investigate friction, observe their ability to work together as a group. Use the cooperative skills teacher assessment sheet on page 21 to record results.
- Have students complete a cooperative skills self-assessment on page 23 to reflect on their own cooperative skills.

Date: ____________ Name: ____________________

Friction

Block	Surface	Force to Pull (N)
Wooden Block	Smooth	
Wooden Block	Carpet	
Wooden Block	Sandpaper	
Wooden Block + Weight	Smooth	
Wooden Block + Weight	Carpet	
Wooden Block + Weight	Sandpaper	

What type of surface increased the friction? ________

____

How do you know the surface increased the friction?

____

____

What type of surface decreased the friction? ________

____

How do you know the surface decreased the friction?

____

____

3 Magnetic Force

Materials

- magnets (different sizes and shapes)
- variety of common objects (e.g., nails, pins, pencils, dime)
- chart paper, felt pens

Activity

Review the definition of force from lesson 1. Explain to the students that when they moved the boxes and knapsack from one spot to another the motion was caused by direct force. In other words, the objects were moved because of an applied force (muscle power). Tell the students that the next three forces they are going to investigate cause motion indirectly. In other words, the force does not touch the object directly to make it move.

Divide the class into small groups and give each group two magnets and a variety of common objects. Explain to the students that they are going to experiment to see which objects are attracted to magnets and which are not. Have them fill in their activity sheet as they experiment. Make sure they list the name of the object and check the appropriate column (i.e., "attracts" or "does not attract"). Remind the students that they will have to conduct the experiment twice (once with each magnet).

As students are experimenting with the magnets, encourage them to see if they can make attracted objects move without actually touching the magnet to the object, by pulling the object with the force of the magnet.

Once the students have filled out their chart, ask them:

- Which objects were attracted to the magnet?
- Which objects were not attracted to the magnet?
- Can you think of any other objects that would or would not be attracted to the magnet?
- Did both magnets work the same way each time?
- Did you have to actually touch the object to move it with the magnet?
- Did the magnet use a pull or a push to move objects?
- What kind of force did you use to move the object with the magnet, direct or indirect?

Now that the students have had an opportunity to experiment with the magnets, ask the students:

- How are magnets helpful in everyday life?
- Where do you see magnets used in everyday life?

Record brainstormed suggestions on chart paper.

Activity Sheet

Note: The activity sheet is to be completed during the activity.

Directions to students:

Record the names of the objects and use a checkmark to record whether or not the objects are attracted to the two magnets (3.3.1).

Extensions

- Encourage students to find other objects around the classroom or at home that are attracted to magnets.
- Have students design experiments to determine if the size of the magnet or shape of the magnet makes a difference in how objects are attracted to it. (Suggestion: See how many paper clips a magnet can pick up. Compare the results with other magnets.)

▶

- Experiment to see if magnets can attract through different materials. Hold a magnet under a piece of paper. Put a metal object (paper clip, thumbtack, nail, and so on) on top of the paper. Move the magnet under the paper and see if the metal object moves. Repeat the activity using different materials (instead of the paper). Suggestions include wood, glass, cloth, and cardboard.
- Brainstorm a list of uses for magnets in everyday life. Have students conduct research on these uses (e.g., compass for navigation, production of electricity).

Activity Centre

Have students draw simple road maps on a sheet of white paper. Now challenge them to make a paper clip follow the road's path using indirect magnetic force (pulling the paper clip along the "road" without actually touching it to the magnet).

Assessment Suggestion

Gather a collection of objects and have individual students sort them according to whether or not the object will be attracted to a magnet. Encourage students to explain their sorting. Also encourage students to provide examples of how magnets are used in everyday life. Use the individual student observation sheet on page 16 to record results.

Date: ____________________ Name: ________________________________

Magnetic Force

Object	Magnet #1		Magnet #2	
	Attracts	Does Not Attract	Attracts	Does Not Attract

4 Static Electrical Force

Materials

- balloons
- pieces of wool, flannel, or felt

Activity

Review with the students the difference between a direct and indirect force (the two previous lessons). Tell the students that they are going to investigate a second force that causes motion indirectly, called *static electricity*.

Note: At this age level, students are not required to use terms such as *atoms, electrons,* and *protons* when discussing static electricity. Rather than focusing on "what is" static electricity, focus on how you can see or produce static electricity.

Keep in mind as well that concepts about static electricity were introduced in the previous unit on Magnetic and Charged Materials. Refer to these activities and encourage students to use their background knowledge to assist them with the following activities.

Divide the class into pairs of students and provide each pair with a balloon and a piece of wool, flannel, or felt. Ask:

- How could you charge the balloon with static electricity?
- What will happen to the balloon when you charge it?

Have each pair charge the balloon by rubbing it with a cloth. Ask:

- What do you think will happen if you move the balloon close to your partner's hair?

Have the students test their predictions and observe what happens when the charged balloon is brought in close proximity to human hair. Ask:

- What kind of force does this show?
- What are some other examples of static electricity?

Review the various activities on static electricity conducted during the previous unit. Discuss investigations with balloons, combs, glass rods, and electroscopes. You may choose to repeat some of these or conduct some of the extension activities for review.

Activity Sheet

Directions to students:

Complete the chart to show your ideas about static electricity. Include a definition, at least three examples of static electricity, and diagrams for each example (3.4.1).

Assessment Suggestion

Review the students' activity sheet charts to determine their understanding of static electrical force. Determine criteria for assessment such as:

1. clear definition
2. labelled diagram
3. example 1
4. example 2
5. example 3

Record these criteria on the rubric on page 19 and record results accordingly.

Date: ____________________ Name: ____________________

Static Electrical Force

Definition	Example	Diagram

5 Gravitational Force

Science Background Information for Teachers

Gravitation is another force that attracts objects to one another. All masses exert gravitational attraction. However, the effect is usually unnoticeable except for very large masses.

We are most familiar with the gravitational attraction of Earth – the force that pulls objects (including humans) toward the centre of Earth. To lift an object off the ground, you must exert an upward force because gravity is pulling the object down. To lift a sheet of paper, you only need to exert a small force. To lift a book, you must exert greater force because the book's mass is greater than the mass of the sheet of paper.

Other masses affect us, too. It is because of the Sun's gravity that Earth orbits the Sun. It is because of the Moon's gravity that ocean tides ebb and flow. On the Moon, gravitational pull is not as great as it is on Earth; the Moon's mass is not as great as Earth's mass. This is why the astronauts who walked on the Moon wore special weighted suits and boots that held them near the Moon's surface. Even then, the astronauts tended to float or bounce when they walked because of the limited gravitational force pulling them downward.

Note: Gravitational force is a rather complex concept and should be introduced to students through simplified activities and discussion. The following activity will assist students in understanding ideas about gravity.

Materials

- ball
- book
- uniform weights (any fairly heavy objects that are compact and easy to tie to a string to act as the pendulum's bob, such as large nuts or washers)
- string
- scissors
- graph paper (included) (3.5.2)

Activity

Review the two indirect forces discussed so far, magnetism and static electricity. Explain to the students that they are going to investigate the third indirect force, called *gravitational force*.

Hold the ball up in the air and ask the students:

- What will happen if I let go of this ball?
- Why will it fall?

Hold the book up in the air and ask the students:

- What will happen if I let go of this book?
- Why will it fall?

Tell the students that they will have an opportunity to see the effects of gravity, using a pendulum. Divide the students into working groups and have them create a pendulum by tying a weight (washer or nut) to the end of a piece of string (50 cm is a good starting length for the string). Introduce the term *bob* to refer to the weight on the end of the string.

Outline the instructions for the investigation: Have one student hold the string steady and the other student release the weight (therefore, there is no need to attach the pendulum to the stand). Explain to students that it is important to release the weight from the same spot each time. (You may wish to introduce 90 degree angles at this time and ensure that students release the pendulum from the point directly in line with the student's hand that is holding the string.) Have the students count the number of times a pendulum swings in thirty seconds.

▶

The third student should be responsible for watching the time. Students should record their findings on their activity sheets.

Once the students have had an opportunity to experiment with the pendulum, ask them:

- Why does the pendulum stop swinging?
- What effect does gravity have on the pendulum?

As a class, brainstorm all the ways to charge the pendulum. Now encourage the students to change the variables and conduct the experiment again. For example, vary the bob's weight by adding more nuts or washers, or vary the string length. Have the students record these changes on the activity sheet, as well as the swings per thirty seconds.

Following the experiment, encourage students to generate some rules for pendulum motion. Ask them:

- What is the effect of lengthening the string? Of shortening the string?
- Does the amount of weight on the string affect the number of times the pendulum swings?

Have the students graph the results of this experiment on the graph paper provided.

Activity Sheet

Directions to students:

Use the chart to record the results of your experiment with pendulums, then use the results to create a bar graph (3.5.1).

Extension

Conduct a visual experiment to observe the motions of the swinging pendulum. Make a pendulum with a slow-pouring funnel at the end of the string. Pour salt or free-flowing sand in the funnel, and place a large piece of black construction paper or mural paper underneath the pendulum on the floor. Have the students note the different patterns formed on the paper as you swing the pendulum. Vary the angle of release and observe changes in the pattern.

Date: ______________ Name: ______________________

Gravitational Force – Swinging Pendulum

Weight (g)	Length of String (cm)	Angle of Release (45° or 90°)	Swings per minute
10 grams	50 cm	90°	
10 grams	50 cm	45°	
20 grams	50 cm	90°	
20 grams	50 cm	45°	
10 grams	25 cm	90°	
10 grams	25 cm	45°	
20 grams	25 cm	90°	
20 grams	25 cm	45°	

Date: ____________________ **Name:** ________________________________

6 Energy and Movement

Materials

- toys that require muscular force for movement, such as toy cars
- toys that store energy to create movement (e.g., jack-in-the-box, elastic band propelled plane, windup toys)
- pencils
- empty thread spools (with a groove cut into the bottom)
- nails
- beads
- elastic bands
- rulers or metre sticks

Activity

As a class, sort the toys into groups according to how they move (by pushing, winding, and so on). Have the students sit in a circle. Place the different toys in the centre of the circle. Select students and ask them to find a way to make each toy move. For example, they can push a toy car, wind up the windup toy, twist the elastic band on the propeller of the airplane to make it fly. Ask the students:

- How did the objects move?
- What did you need to do to make the objects move?

Explain to the students that they are going to make their own elastic band motor. Divide the class into working groups. Give each group the required materials. You may wish to provide each group with a separate sheet of instructions, write the instructions on chart paper, or work through the instructions orally.

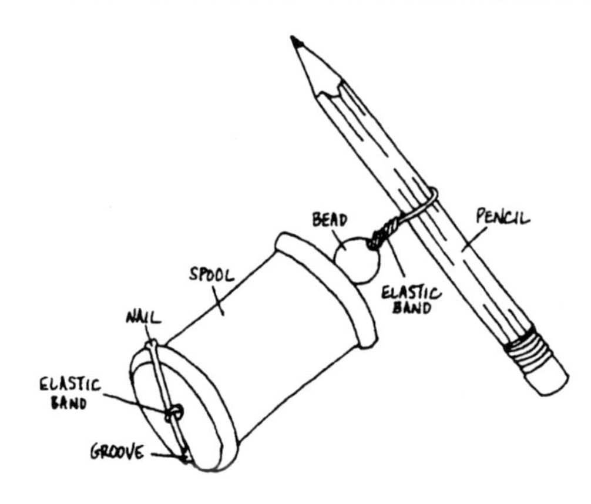

Directions for making the elastic band motor.

1. Thread an elastic band through the centre hole of the spool.
2. Put a nail through the elastic band and fix it into the groove.
3. Thread the other end of the elastic band through a large bead.
4. Fix a pencil to the end of the elastic band.

Explain to the students that they are going to wind the pencil, then measure how far the spool travels. It will depend on the number of times they wind the pencil. Have students record their findings on their activity sheet.

Activity Sheet

Note: The activity sheet is to be completed during the activity.

Directions to students:

Use the chart to record the movement of your elastic band motor (3.6.1).

Activity Centre

Have students bring in other toys that can be moved using muscular energy or stored energy. Provide students with opportunities to use force to create movement. Have them discuss their investigations using appropriate vocabulary such as *force, push, pull, direction, distance,* and *speed.*

6

Extension

Bug in the Envelope: Bend a bobby pin into a V-shape. Thread a button onto an elastic band and stretch the elastic band over the two ends of the bobby pin (this is the "bug"). Wind up the button by twisting it on the elastic band. Place the button in an envelope, ensuring that the button stays wound up. When the envelope is opened, energy will be released, causing the "bug" to move.

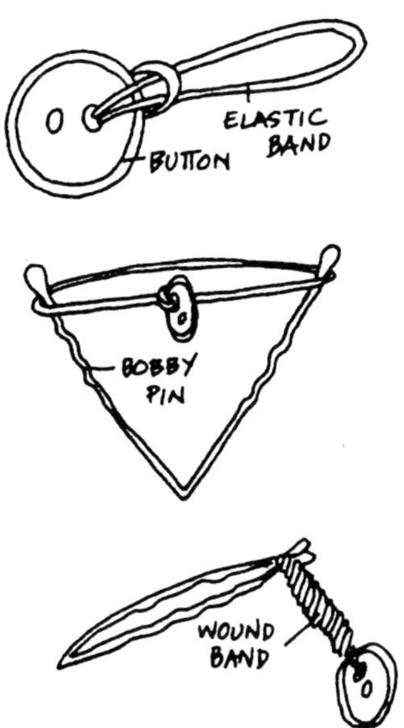

Date: ________________ Name: ______________________________

Elastic Band Power

Number of Winds	Distance Travelled	Observations
10		
20		
30		
40		
50		

Explain in your own words how the spool travelled. Where did it get its energy from? ______________________

____

____

____

7 Designing Toys That Use Different Forms of Energy to Move

Materials

Note: In this activity, students are going to design and construct their own toys using one of the four forces of movement (muscular force, magnetism, static electricity, gravitational force); therefore, materials will vary with each student. You may wish to provide an array of materials that students can build their models from, and/or encourage students to bring materials from their own home. The book *Wind-Ups* by Chris Ollerenshaw and Pat Triggs provides several ideas for toy designs.

Activity

Explain to the students that they are going to work in pairs or by themselves to design, construct, and present a toy that moves using one of the four forces of movement. Review the four forces with the students. Record the name of each of the four forces on chart paper and give examples of each force.

Have the students brainstorm ideas of toys they could build. Record these suggestions on chart paper. You may wish to show the book *Wind-Ups* to the students. It is filled with coloured photos, directions, and examples of homemade toys.

Outline the project for the students.

1. Decide whether you are going to work by yourself or with a partner.
2. Create a list of different toys you would like to make.
3. Select one toy from the list of suggestions.
4. Identify and record the force used to move the toy (e.g., muscular force, gravitational force, magnetism, static electricity).
5. Draw a labelled diagram of your toy.
6. Make a list of the materials you will need to build your toy.
7. Identify where you can obtain the materials required.
8. Gather the necessary materials.
9. Construct the toy, using your diagram as a guide.
10. Write a paragraph describing your toy. Make sure you include the name of the toy, what force is needed to move the toy, what is special about the toy, and how you constructed the toy.
11. Present your toy to the class.

Activity Sheet

Note: The activity sheet is to be completed during the planning, design, and construction phases of the activity.

Directions to students:

Work through the plan for designing and constructing your toy (3.7.1).

Extensions

- Set up a display of the different toys the students created as well as their planning guides. Make sure the display is available to other classes of students by locating it in the school library, main hallway, or outside your classroom.
- Have the students visit other students in the school to describe their toy, how it is used, and how they constructed it.
- Invite a guest speaker from a local toy store or toy company. Have the speaker show a variety of toys and the force used to make each move.
- Create a big class book called *Our Designs*. Have each student contribute a page that includes a diagram and a description of the toy he or she constructed.
- Have students pretend they work for a toy manufacturer. Have them promote and advertise their toy as the toy of choice.

7

Assessment Suggestions

- As a class, develop criteria for the toy construction project by which students will be evaluated. Display these criteria for students to reference throughout the project. Record these criteria on the rubric on page 19. When students make their final presentations, the rubric can be completed.
- Have students complete the student self-assessment on page 22 to reflect on their own learning during this project.

Date: ________________ Name: ______________________________

____

Toy Designs

List of suggested toys:

______________________ ______________________

______________________ ______________________

______________________ ______________________

______________________ ______________________

I/We have selected the following toy to design and build: ______________________________

This toy will use the following force: ________________

____

Materials Required:

______________________ ______________________

______________________ ______________________

______________________ ______________________

______________________ ______________________

______________________ ______________________

Date: ____________ Name: ____________

Diagrams of Toy

Front View	Top View	Side View

Description of Toy: ____________

8 Everyday Devices

Materials

- chart paper, felt pens
- pictures of everyday devices (from magazines and newspaper flyers)
- glue
- scissors
- mural paper

Activity

Begin the lesson by turning the classroom lights off and on. Ask the students:

- What device did I just use?
- What happened when I switched the light switches?
- What kind of force did I use to turn the lights off and on?

Explain to the students that many things in our everyday life function/work as a result of force. Some devices are controlled automatically, some are controlled at a distance, and others are controlled by hand. Divide chart paper into three sections. Title the columns Controlled by Hand, Controlled Automatically, and Controlled at a Distance.

First, have the students brainstorm different devices in the home, school, and community that function because they are controlled by hand. These may include light switches, door knobs, toilet handles, car ignitions, water taps, and hand-held hose nozzles. Record these ideas on the chart.

Discuss devices that are controlled automatically and do not require actual hand force. These might include school bells, light timers, dishwashers, washing machines, dryers, and traffic lights. Record these ideas on the second column of the chart.

Devices controlled at a distance include remote-control toys, televisions (using a remote control), and garage door openers. Discuss these devices and record the students' ideas on the third column of the chart.

Once the students have an understanding of each category, divide the class into working groups. Provide each group with magazines, catalogues, flyers, scissors, and glue. Have the students look through the magazines and flyers to find pictures of everyday devices that fall into each of the three categories.

Divide a large sheet of mural paper into three sections and label each section according to the categories on the chart. Have the students glue their pictures onto the appropriate sections of the mural. Display the mural along with the chart.

Activity Sheet

Note: The activity sheet is to be completed by students at home as a follow-up to the activity.

Directions to students:

Go around your home and make a list of as many devices as you can find that fit into the three categories listed on the chart (3.8.1).

Date: ______________ Name: ______________________

Everyday Devices

Controlled Automatically	Controlled at a Distance	Controlled by Hand

9 Parts of a System

Materials

- chart and mural from previous lesson
- toaster
- chart paper
- rulers
- pencils
- erasers
- pencil crayons or markers
- magazines and catalogues displaying household appliances and devices

Activity

Using the chart and mural, review the devices identified in the previous lesson (devices controlled automatically, devices controlled at a distance, devices controlled by hand). Explain to the students that they are going to work together to examine the different parts of one device, specifically how the parts work together to perform a specific function. Before you have the students do this, complete an example project with the whole class. Display a toaster for students to examine. Discuss its parts and explain how it works.

1. The toaster plug in the outlet takes electricity from the outlet to the cord.
2. The cord sends electricity to the coils in the toaster so that they can heat up.
3. The lever on the side of the toaster is pushed down (muscular force/direct motion) to drop the bread into the toaster.
4. The coils in the toaster heat up (turn red) to toast the bread.
5. Temperature detectors inside the toaster shut off when the coils reach the desired temperature.
6. The toaster lever springs up when the coils shut off.

Ask the students:

- What would happen if the toaster did not have a plug?
- What would happen if the toaster did not have the temperature detector inside?
- What do you know about the different parts of a system, in this case a toaster?
- What happens if one of the parts of a system is missing?

Explain to the students that they are going to look at the parts of the device they have chosen. Divide the class into working groups. Have the groups cut out a picture of a device from a magazine, catalogue, or, if possible, collect several devices so students can examine them firsthand.

Note: If students are handling household devices, ensure that, for safety reasons, the activity is supervised.

Have each group draw a labelled diagram of the device on the activity sheet, then explain in specific steps how the different parts of the device work together to perform a specific function.

Have each of the groups present its diagram and description to the other members of the class.

Activity Sheet

Directions to the students:

Draw a diagram of your selected device and label each part. Explain how the parts work together so that the device performs a specific function (3.9.1).

▶

9

Extensions

- Visit a local food processing plant so that students can view firsthand the different parts of the system working together to make a final product.
- Investigate different systems around the school (e.g., the PA system). Have the students break the system down into its parts and discuss how the parts work together to perform a specific function.

Activity Centre

Create a "take-apart" centre: collect several household items for students to examine and take apart. Include old watches, clocks, radios, calculators, and so on along with screwdrivers. Allow students to examine the internal workings in order to determine how the parts work together to perform a specific function.

Date: ____________ Name: ____________________

____

____

____

Looking at the Parts of a System: Household Devices

Name of household device: ____________________

References for Teachers

Ardley, Neil. *The Science Book Of Gravity*. New York: Gulliver Books, 1992.

Challand, Helen. *Experiments With Magnets*. Chicago: Children's Press, 1986.

Dixon, Malcolm, and Karen Smith. *Forces & Movement*. London: Evans, Brothers, Ltd., 1997.

Fowler, Allan. *What Magnets Can Do*. Chicago: Children's Press, 1995.

Gibson, Gary. *Playing With Magnets*. Brookfield: Copper Beech Books, 1995.

Hall, Godfrey. *Back to Basic Science for 8-9 Year Olds*. London: Letts Educational, 1996.

Jennings, Terry. *Magnets*. New York: Gloucester Press, 1990.

Merrell, JoAnn. *Force and Motion*. Huntington Beach, CA: Teacher Created Materials, 1994.

Ollerenshaw, Chris, and Pat Triggs. *Wind-Ups*. Milwaukee: Gareth Stevens Publishing, 1991.

Parker, Steven. *Magnets*. Milwaukee: Gareth Stevens Publishing, 1991.

Peacock, Graham. *Electricity*. East Sussex: Wayland Publishers, 1993.

Pressling, Robert. *My Magnet*. Milwaukee: Gareth Stevens Publishing, 1994.

Royston, Angela. *What's Inside? Toys*. Toronto: Grolier, 1991.

Taylor, Barbara. *Force & Movement*. Toronto: Franklin Watts, 1990.

Ticotsky, Alan. *Who Says You Can't Teach Science?* Glenview: Scott, Foresman and Company, 1985.

White, Laurence. *Gravity–Simple Experiments for Young Scientists*. Brookfield: Millbrook Press, 1995.

Whyman, Kathryn. *Forces In Action*. Toronto: Gloucester Press, 1986.

Unit 4

Stability

Books for Children

Gibbons, Gail. *Up Goes the Skyscraper*. New York: Aladdin Books, 1990.

Hughes, Shirley. *The Big Cement Mixer*. Toronto: Douglas & McIntyre, 1990.

Kellogg, Steven. *The Three Little Pigs*. New York: Morrow Junior Books, 1997.

Knight, Margy Burns. *Talking Walls*. Kingston, ON: Quarry Press, 1992.

Pfanner, Louise. *Louise Builds a House*. New York: Orchard, 1989.

Robbins, Ken. *Building a House*. New York: Four Winds Press, 1984.

Steltzer, Ulli. *Building an Igloo*. New York: H. Holt, 1995.

Thompson, Colin. *Tower to the Sun*. New York: Knopf, 1997.

Van Allsburg, Chris. *Ben's Dream*. Boston: Houghton Mifflin, 1982.

Wilcox, Charlotte. *A Skyscrapers Story*. Minneapolis: Carolrhoda Book, 1990.

Wilkinson, Philip. *Amazing Buildings*. New York: Dorling Kindersley, 1993.

Zelinsky, Paul. *The Maid, the Mouse and the Odd-Shaped House*. New York: Dodd, Mead & Company, 1981.

Web Sites

- **http://www.pbs.org/wgbh/nova/bridge**

  NOVA: discover the program "Super Bridge" and find out how to plan and construct four bridges on your own. This site offers great teacher resources, lesson ideas, and "Build a Bridge" activities.

- **http://www.iti.acns.nwu.edu/clear/bridge**

  A helpful teacher resource site on bridges. Provides web sites, articles, and a large section on special topics from "bridge-coating" to "bridge testing."

- **http://www.struct.Kth.se/people/raid/cable.htm**

  This site about "Cable Supported Structures" offers numerous pictures and descriptions about cable-stayed, suspension, and the world's largest bridges.

- **http://www.xs4all.nl/~hnetten/tallest.html**

  Find out which structures in the world are really the tallest! Discover detailed information about many different towers from around the world. Also included are maps and extensive information about the countries that have built the structures.

- **http://www.greatbuildings.com/**

  "The Great Buildings Collection" is a gateway to architecture around the world and across history. An excellent resource that gives background information about famous architects, buildings, and structures.

- **http://www.nmsi.ac.uk/on-line/challeng/index.html**

  The Science Museum of London, England: "Challenge of Materials" series is an excellent site for teachers and students. Learn about how materials are made, their molecular structure, how materials are selected, and materials that have changed our lives. Great graphics and pictures make the information accessible.

- **http://www.pbs.org/flw/**

  PBS: a detailed look at one of North America's greatest architects, Frank Lloyd Wright. Look into this famous man's life and work through his drawings, interiors, exteriors, and critical responses to his work.

- **http://www.apase.bc.ca/**

  Association for the Promotion and Advancement of Science Education: an excellent resource for teachers and students. Click on "Charlotte's Web" for educational ideas on human environments – includes architecture for kids, rhythm and shape, urban design, and much more.

- **http://library.advanced.org/18788/**

  Architecture Through the Ages: an educational site that discusses all types of architecture from around the world and time periods. Excellent site for teachers and students researching architecture.

- **http://www.endex.com/gf/buildings/itpisa.html**

  Leaning Tower of Pisa: scroll down for history and information on the town of Pisa and its architecture (including the Tower of Pisa). Includes photographs and other visuals.

Introduction

In this unit, students will develop their understanding of the concept of stability in structures and the functions of various mechanisms. They will design and build strong and rigid structures, reinforcing the concepts that have been introduced. Students will also gain some understanding of balance, through investigations that focus on balance points and locating the centre of gravity. This will provide the necessary foundation for the study of equilibrium.

By the end of this unit, students will demonstrate an understanding of the factors that affect the stability of objects. They will design and make structures that include mechanisms that can support and move a load, and investigate the forces acting on them. Students will also describe, using their observations, systems involving mechanisms and structures, and explain how these systems meet specific needs and how they have been made.

Science Vocabulary

Throughout this unit, teachers should use, and encourage students to use, vocabulary such as: *strength, force, function, lever, fulcrum, load, effort, balance, balance point, centre of gravity, beam bridge, arch bridge, suspension bridge, movable bridge, truss, beam, strut, girder, tie, cable, structural,* and *mechanical.*

Materials Required for the Unit

Classroom: paper (21.5 cm x 28 cm), pencils, hardcover books, tape, 30-cm rulers, masking tape, erasers, red and blue crayons (or markers or pencil crayons), rulers, metre sticks, scissors, Plasticine, Manila tag (or Bristol board), books (for weights), chairs, string, glue sticks, glue, paper fasteners, single-hole punch

Books, Pictures, and Illustrations: pictures of common levers, picture of seesaw (included), books and pictures that illustrate a variety of different bridges, reference materials

Household: forks, spoons

Equipment: computer (and access to the Internet), low-temperature glue guns

Other: small uniform weights, common levers (e.g., crowbar, hammer, scissors, pliers, wheelbarrow, tongs), pennies, balls, Hula-Hoops, balance beam, large cardboard box, packing tape, nails, straws, toy cars, toy airplanes, toy trucks

1 Altering the Strength of Objects

Materials

- sheets of paper, 21.5 cm x 28 cm
- small uniform weights

Note: If you do not have access to weights you can use other uniform items such as pennies, nuts, or washers.

- pencils
- hardcover books
- tape

Activity

Divide the class into working groups of two or three. Give each group paper, tape, a stack of books, some weights, and activity sheets.

Have the students place two piles of books on a flat surface so that each pile is the same height and approximately 20 cm apart. Place a piece of paper over the gap so it spans the two piles of books. Ask:

- How many weights do you think the paper can hold before it collapses?

Have the groups test their predictions by placing weights on the paper until it buckles. Ask:

- What can you do to the paper so that it can support more weight?

Have the students experiment and try to make the strongest paper bridge using up to three sheets of paper and tape.

Note: Tape can be used to join the pieces of paper, but it can not be used to tape the paper to the books.

Have the students predict and record the maximum number of weights their best paper bridge could support, then draw a diagram of the bridge on their activity sheets in the section titled Trial #1.

Once all of the groups have had an opportunity to complete the initial experiment, have each group hold up its paper bridge and tell how many weights the paper bridge was able to support. Ask the students:

- Why were some paper bridges able to hold more weights than others?
- What was different about the paper bridges?
- What does folding a piece of paper do to the strength of the piece of paper?
- Can you make changes to your paper bridge so that it holds more weights?

Note: If none of the groups thinks of folding the paper to increase the strength, you may wish to ask some additional probing questions or show an example. Sample questions:

- How could you change the shape of your piece of paper?
- Would changing the shape of your piece of paper alter its strength?
- In what different ways could you fold the paper? How would this affect the strength of your bridge?

Have students conduct the experiment once again, altering the shape of their paper bridge by folding it a different way, so that it can hold more weights. Have students record the number of weights their paper bridge can hold in this second trial. In addition, have them draw a diagram of their paper bridge.

Following this activity, discuss ways in which forces alter the shape or strength of different structures. Ask the students:

- What happened to the paper bridge when you added more weight to it?
- Did its shape change? How?
- What happened when the paper bridge could no longer support the force of the weight?

▶

1

Explain to the students that a force is a push or a pull. (This concept is also introduced in the previous unit titled, Forces and Movement.) When weights were added to the paper bridges, they pushed down on the paper. This force caused the paper to collapse. Ask:

- Can you think of other examples to show how the force of weights can alter the shape or strength of an object?

Note: Look for examples around the classroom. These may include a bookshelf that is sagging due to a heavy load of books, a cardboard box that buckles when too much weight is placed inside, or a paper or plastic bag that breaks when the load inside is too heavy.

Activity Sheet

Note: The activity sheet is to be completed during the activity.

Directions to students:

After making a paper bridge, predict and record the number of weights your bridge can hold. Draw a diagram of your bridge design. Do the same for your second bridge (4.1.1).

Extensions

- Give students opportunities to examine other ways that forces alter the shape or strength of different structures. Some questions you might ask are:
  - How many books can you place in a box before the box buckles?
  - How many weights can you hang from a thread before the thread breaks?
  - What happens when books are placed on a Plasticine tower?
  - How many marbles can you place in a paper lunch bag before the bag tears?

  Encourage students to devise their own investigations and record their results.
- Examine how force is used to make specific mechanisms work. On chart paper, record tasks such as:
  - flushing a toilet
  - turning on a tap
  - opening a door
  - sharpening a pencil

  Ask the students:
  - How do you flush a toilet (turn a tap on/open a door/sharpen a pencil)?
  - For each of these mechanisms to work, what does each need? (A force acted upon it.)

  Have the students brainstorm other structures, mechanisms, or objects in the classroom that require a force to function or perform a specific task. Record their suggestions on the chart paper.

Activity Centre

Provide the following materials at the centre:

- thin sheets of cardboard or Bristol board, cut into small squares (approximately 10 cm x 10 cm)
- white glue
- gram weights
- books

Have the students pile two sets of books and place a piece of cardboard on the books so that it spans the two piles. Challenge the students to predict and test the strength of the cardboard by placing weights on it until it buckles. Now have the students glue two pieces of the cardboard together, let dry, and test for strength. Continue with three, four, and five sheets of cardboard glued together to determine how the number of layers affects the strength of the structure. Students can create their own charts to record results.

Date: ______________ **Name:** ______________________

____

____

____

Paper Bridges

Trial #1 - Our Bridge Design

Number of weights the paper can support:

Prediction______________ **Result**______________

Trial #2 - Our Bridge Design

Number of weights the paper can support:

Prediction______________ **Result**______________

2 Levers

Science Background Information for Teachers

A lever is a rigid bar that rests on a fixed, pivotal point called the fulcrum. The lever helps make work easier through the use of force and distance. There are three types of levers – first class, second class, and third class. The location of the fulcrum with relation to the load and the effort determines the class of lever.

First class lever: The fulcrum is between the effort and the load. Examples of a first class lever are scissors and a seesaw.

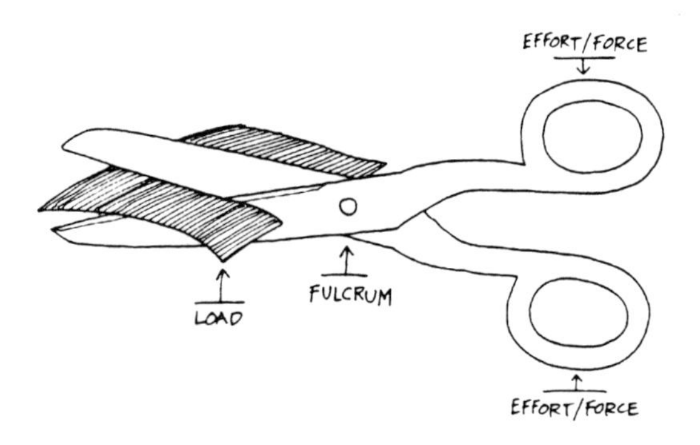

Second class lever: The fulcrum is at one end and the load is between the fulcrum and the effort. A wheelbarrow is an example of a second class lever.

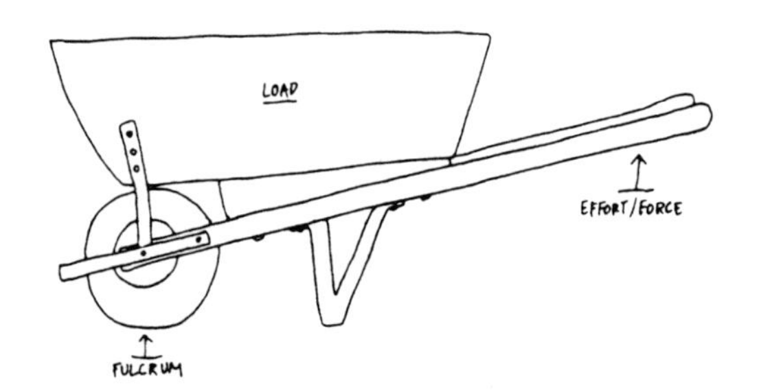

Third class lever: The fulcrum is at one end and the effort is between the fulcrum and the load. Tweezers and tongs are examples of third class levers.

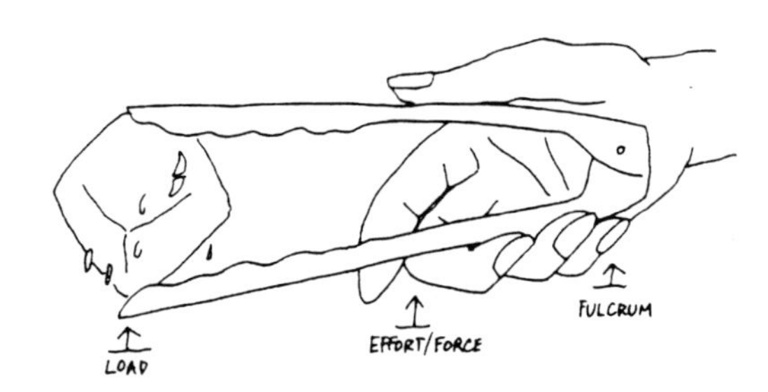

The closer the fulcrum is to the load, the less the amount of force that is needed to lift the load.

Materials

- variety of common levers (or pictures of common levers) such as a crowbar, scissors, hammers, pliers, wheelbarrows, tweezers, and tongs
- picture of a seesaw or teetertotter (included) (4.2.1)
- pencils
- 30-cm rulers
- pennies
- masking tape

Activity

Display a variety of levers. Tell the students that all of these objects have something in common. Allow them time to explore the objects. As they explore, challenge them to identify what the objects have in common.

Once the students have had time to examine the levers, ask:

- What are the names of these objects?
- How are they each used?
- What is the same about how they work?

Explain to the students that the objects in this collection make work easier for humans. Review the examples and discuss how each makes work easier. Tell the students that these items are called *levers*. Identify for the students

▶

the three components of a lever (force or effort, load, fulcrum).

Use the seesaw illustration (4.2.1) as an example of a first class lever. The board is supported by a stand called the *fulcrum*. A person who sits at one end of the seesaw is the *load*. The person that sits at the other end is the *force* or *effort* that is used to lift the load.

Ask the students:

- Can an adult and a child play on a seesaw together?
- Can the child lift the adult?
- What can be done to help the child lift the adult?

Explain to the students that they are going to conduct their own seesaw experiment. Divide the class into working groups and provide each group with a ruler, a pencil, pennies, masking tape, and an activity sheet. Have the groups construct a model seesaw. (Place the ruler on top of the pencil. The pencil acts as the fulcrum, allowing the ends of the ruler to move up and down.) Have the students tape five pennies to one end of the ruler. The pennies will act as the load to be lifted on the seesaw.

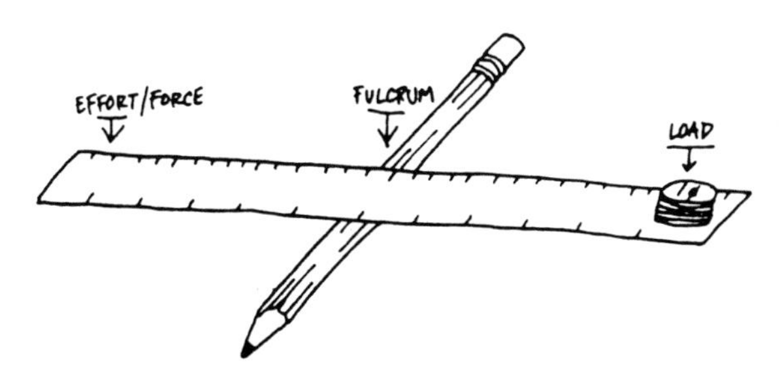

Give the students some time to manipulate the materials, lifting the load by pushing down on the other end of the ruler, and moving the fulcrum. Ask:

- Do you think that changing the position of the fulcrum affects the amount of force required to lift the load?

Guide the students as they begin the experiment. Have the groups place the fulcrum at the 15-cm mark on the ruler. Ask:

- What is the distance from the fulcrum to the load?

Explain that they are going to add pennies to the other end of the ruler until the load is lifted. This is called the force that lifts the load. Ask:

- What is the distance from the fulcrum to the force?

Have the students record this measurement on the activity sheet. Ask:

- How many pennies do you think you will need to lift the load?

Have the students record their predictions, conduct the experiment, and record the number of pennies needed to lift the load.

Continue the experiment, changing the position of the fulcrum. With each change, have the students predict and record the number of pennies that were needed to lift the load.

Once students have completed the investigation, ask:

- If the fulcrum is moved closer to the load, how does this affect the amount of force required to lift the load?
- If the fulcrum is moved farther away from the load, how does this affect the amount of force required to lift the load?
- What conclusions can you draw?

2

Activity Sheet

Note: The activity sheet is to be completed during the activity.

Directions to students:

Complete the chart for your seesaw experiment, then draw a labelled diagram of your lever (4.2.2).

Extensions

- Have the students write up this investigation as an experiment, using the extension activity sheet provided (4.2.3).
- Have students identify the different parts of a lever (fulcrum, force, load) using the collection of common levers.
- Investigate and distinguish among first class levers, second class levers, third class levers, and pairs of levers. Conduct similar investigations to the one above using different classes of levers.
- Investigate how marionettes use levers to operate the limbs of the puppet. Have students create their own marionette or simple puppet.

Assessment Suggestion

Observe students as they investigate levers. Focus on their ability to:

- follow instructions
- manipulate materials
- measure accurately
- make predictions
- record results

List these criteria on the rubric on page 19 and record observations.

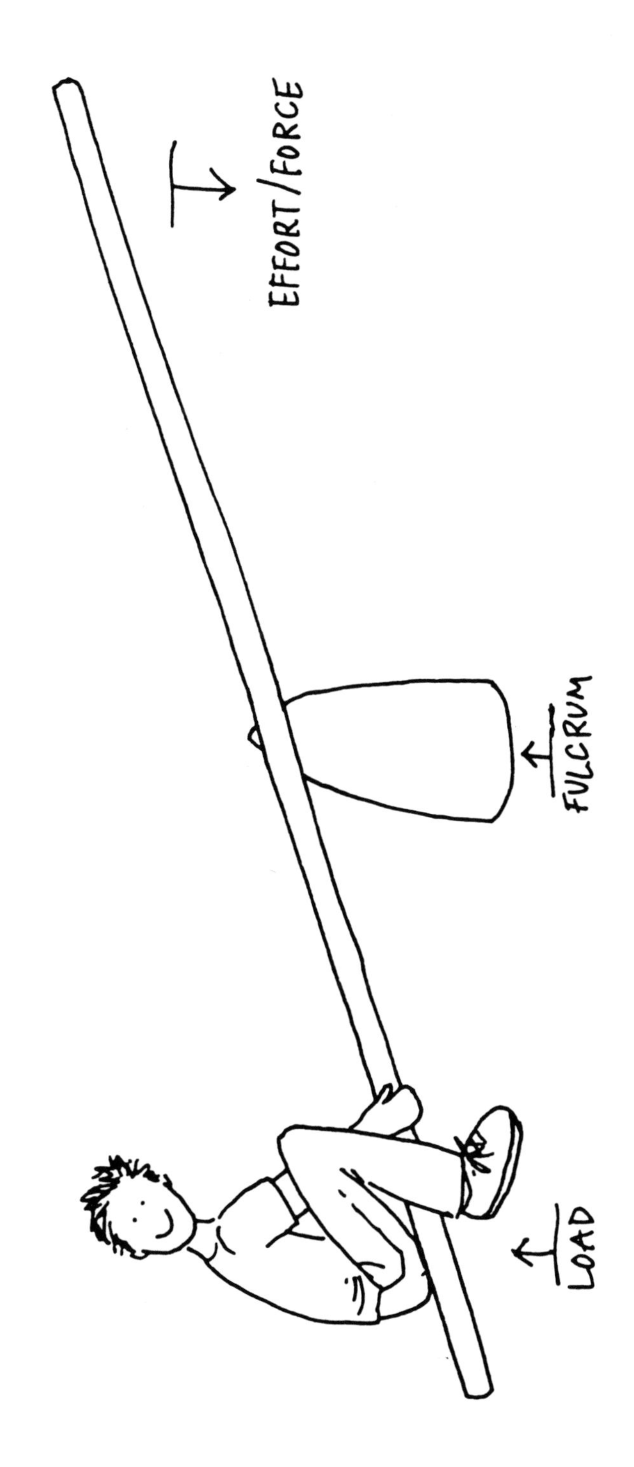
EFFORT/FORCE
FULCRUM
LOAD

Date: ______________ Name: ______________________

____

____

____

Levers

Number of Pennies in Load	Distance from Fulcrum to the Load (cm)	Distance from Fulcrum to the Force (cm)	Force (number of pennies) Needed to Lift Load	
			Prediction	Result
5	15 cm	15 cm		
5	5 cm			
5	10 cm			
5	20 cm			
5	25 cm			

Diagram of our Lever:

Date: ____________ Name: ____________

Experimenting With Levers

Purpose (what you want to find out): ____________

____

____

Hypothesis (what you think will happen): ____________

____

____

Materials (what you need): ____________

____

____

Method (what you did): ____________

____

____

____

____

____

Results (what happened): ____________

____

____

Conclusion (what you learned): ____________

____

____

Application (how you can use what you learned):

____

____

____

3 Balance Points

Materials

- pencils
- erasers
- red and blue crayons, markers, or pencil crayons
- rulers
- metre sticks
- other objects to balance (e.g., fork, spoon, scissors, balls, Hula-Hoops)
- balance beam
- Plasticine

Activity: Part One

As a class, introduce the concept of balance. Ask:

- What does the word *balance* mean?
- Can you balance on one foot?
- What helps you to balance on one foot?

Provide all students with a pencil and challenge them to try to balance it on a finger. Ask the students:

- How were you able to get the pencil to balance on your finger?
- Why does the pencil balance?

Explain to the students that the point at which the pencil rests in a level position is called its "centre of gravity." Tell the students that every object has a point around which all its weight seems to be centered. Once the centre of gravity is found, an object can be balanced. In this case, the student's finger is actually acting as a fulcrum, just like the fulcrum on the levers they experimented with in the previous activity.

Have the students try to find the centre of gravity for other objects (e.g., ruler, metre stick).

When the students have had an opportunity to find the balance points of other objects, put them up to another challenge. Give each student a small ball of Plasticine. Have them push one end of the pencil into the Plasticine. Have students hypothesize where they think the balance point will be. Have them draw a diagram of the pencil with the Plasticine ball attached at one end and record their prediction on their activity sheet. Now have the students balance the pencil with the Plasticine ball on a finger. Once they have located the balance point, have them record their actual findings (centre of gravity) on their activity sheet.

Have the students experiment with other items to see if they can locate the balance point (centre of gravity). Have them record their predictions and actual findings on their activity sheets.

Activity Sheet

Note: The activity sheet is to be completed during Activity: Part One.

Directions to students:

Answer the questions at the top of the sheet (4.3.1).

Activity: Part Two

In a large area such as the gym, challenge students to balance under different conditions. For example:

- stand on one foot
- do a headstand
- walk on a balance beam
- balance a ball on one finger
- balance a Hula-Hoop in the palm of the hand

As the students work through these activities, discuss how different body positions help to provide more stability for balancing themselves or other objects (e.g., spreading out the arms when balancing on one foot or walking on a balance beam, positioning the head and hands in a triangle-shape to do a head stand).

3

Extensions

- Using balance scales, have students attempt to balance different objects (e.g., place a potato on one pan and attempt to balance the scale by placing pennies, washers, or weights on the other pan). This activity can be extended to focus on standard mass measurement by estimating and recording the mass of various objects in grams using the balance scales.
- Investigate how balance affects the stability of a structure. Have the students build simple paper towers using a sheet of paper taped into a cylindrical shape. Place a paper plate on top of the cylinder. Have the students place weights, pennies, or washers on the paper plate until the cylinder collapses. As they conduct this investigation, have them experiment with placing all the weights on one side of the paper plate. Since the plate will not be balanced, the structure will collapse. When the weights are spread out evenly around the plate, more can be added before the structure collapses.

Date: ____________ Name: ____________________

Balance Points

1. Draw a picture of the item to be balanced.
2. Predict where you think its balanced point is located (draw a blue arrow to indicate this point).
3. Find the balance point, and draw a red arrow to indicate the actual balanced point.
4. Experiment with other items to find the balance point.

4 Types of Bridge Structures

Materials

- books and pictures that illustrate a variety of different bridges (see references for suggestions)
- reference material and access to the Internet for research on bridges

Activity

Introduce the topic of bridges by reviewing one of the books that shows a variety of bridge structures. Ask:

- What kinds of bridges do you see?
- How are the bridges the same?
- How are the bridges different?
- What materials are the various bridges built from?
- What do we use bridges for?

Explain to the students that they will be working in pairs to conduct research on one of these four different types of bridges:

- beam bridge
- arch bridge
- suspension bridge
- movable bridge

Review the research steps with the students. For each type of bridge:

1. Use the resources provided to find a picture(s) of the bridge.
2. Draw a diagram of the bridge in the space provided on the activity sheet.
3. Describe the bridge in your own words.
4. Answer the questions at the bottom of the activity sheet.

Note: Ensure that all four types of bridges are focused on by the class as a whole. Once the students have completed their research, they can share their findings with the class and learn about the other types of bridges through discussion and pictures.

Following the research, ask the students:

- What does a (beam, arch, suspension, movable) bridge look like?
- Do we have this type of bridge in or near our community?
- What is special about the design of this bridge?
- What do you think gives this bridge its strength?

Activity Sheet

Note: The activity sheet is to be completed during the activity.

Directions to students:

Research one type of bridge. Include a diagram of the bridge and a description of the bridge. Answer the questions at the bottom of the page (4.4.1).

Extension

Have students conduct further research on famous bridges around the world (e.g., Tower Bridge (London), Bridge of Sighs (Italy), Mackinac Bridge (Michigan), Camel-Back Bridge (China), Bayonne Bridge (New York – New Jersey), "Queen Emma" Pontoon Bridge (Netherlands), Royal Gorge Bridge (Colorado), Verrazano-Narrows Bridge (New York), Rainbow Bridge (Utah), Golden Gate Bridge (San Francisco), Lions Gate Bridge (Vancouver – North Vancouver), Confederate Bridge (N.B. – P.E.I.). Students should illustrate a picture of their famous bridge and describe the bridge. Collate these research pages together and make a class big book called *Famous Bridges Around the World*.

4

Note: The book *Ben's Dream* by Chris Van Allsburg can be used as a pattern for the class book on famous bridges. In the book, Ben's dream takes him around the world to famous landmarks. The class book can follow a similar trip around the world, showing famous bridges.

Assessment Suggestions

- Observe the students as they work together to conduct their research. Use the cooperative skills teacher assessment sheet on page 21 to record results.
- Also have the students complete a cooperative skills self-assessment on page 23 to reflect on their ability to work together.

Date: ______________ Name: ______________________

____

Looking at Bridge Structures

Type of Bridge: ______________________

Diagram:

Description: ______________________

____

____

____

____

What is special about the design of this bridge? ______

____

____

____

What do you think gives this bridge its strength? ______

____

____

____

5 Bridge Construction

Science Background Information for Teachers

Bridges include a variety of components that help to strengthen and stabilize the structure:

triangulation (trusses): used for horizontal bracing of a bridge. Triangles are organized in shapes called trusses. Trusses add strength to a bridge. Structural engineers use knowledge of geometry and stable shapes that will contribute to strength and stability of a structure.

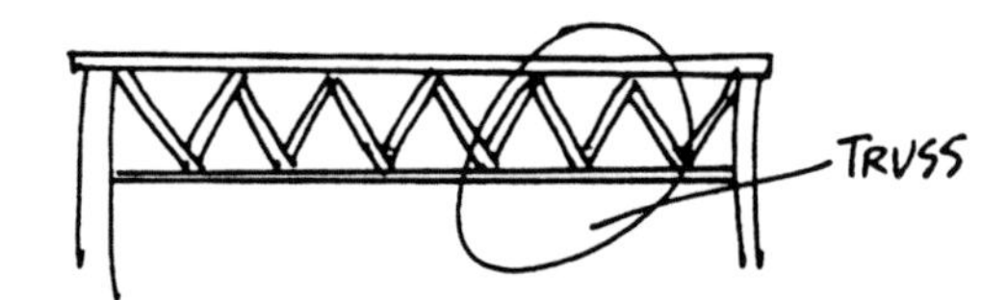

struts: bars that give support or strength to a structure through compression

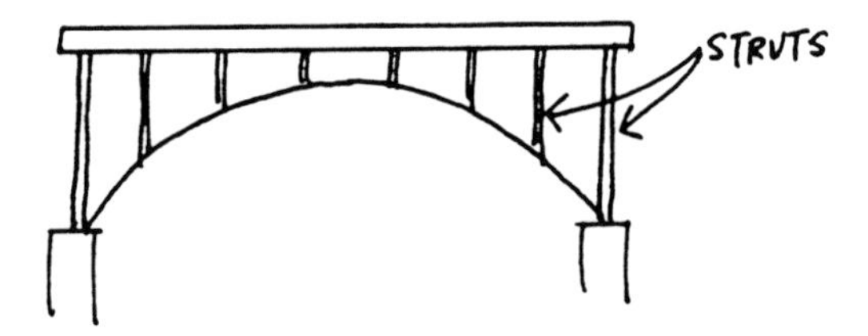

beams: long pieces of wood, concrete, or steel that support the roadway of a bridge

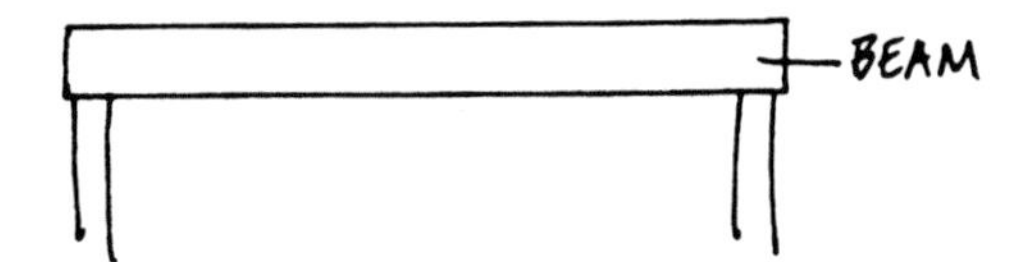

girders: horizontal iron or steel beams that support a load

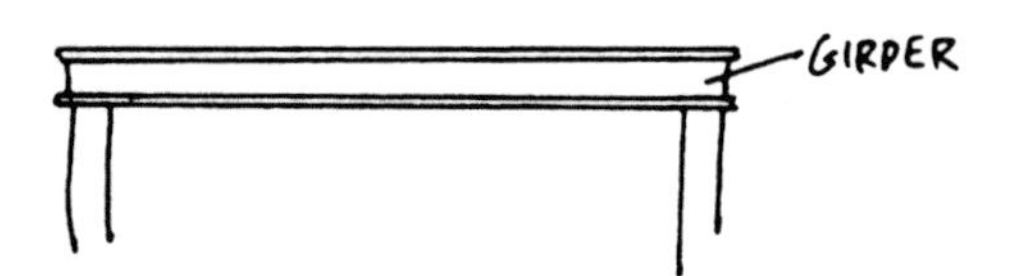

ties: joins two structural members together through tension

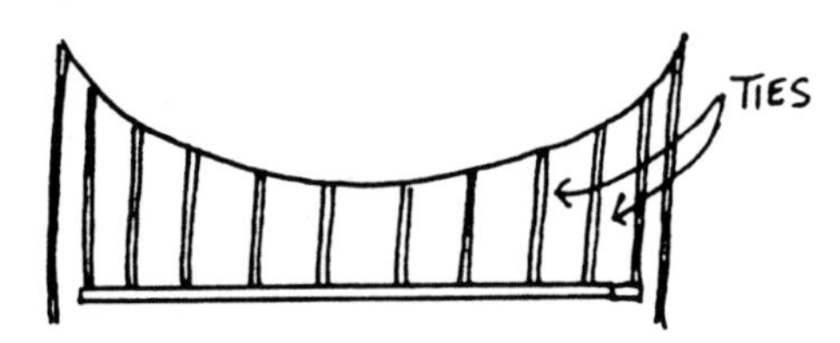

cables: flexible steel ropes that support the roadway of a suspension bridge

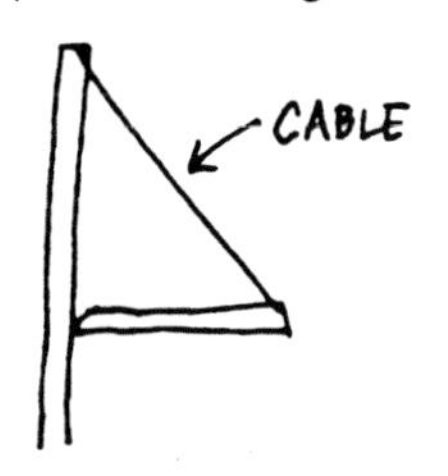

Materials

- books and pictures that illustrate a variety of different bridges (see references for suggestions)
- activity sheet from previous lesson (research sheets)
- chart paper, felt pens
- various materials for constructing bridges as outlined in Activity: Part Two
- illustrations of "bridges" (included) (4.5.1-4.5.4)

Activity: Part One

Have the students review their research sheets that they completed in the previous lesson. Hold up a picture of each type of bridge (beam, arch, suspension, movable) studied. Once again, for each bridge, ask:

- What type of bridge is this?
- What is special about the design of this bridge?
- What do you think gives this bridge its strength?

5

As you review each bridge, sketch a diagram of the bridge on chart paper and introduce the proper vocabulary for the different parts of the bridge structure, as well as the function of each part. Have the students identify and label those parts found on the bridge diagrams they drew as part of their research.

Activity: Part Two

Focus on the different types of bridges and their strucural components. Introduce the four activities (A,B,C, and D) and write the instructions for each on chart paper for the students to refer to as they investigate bridges. As a class, examine the diagrams of the bridge structures (4.5.1-4.5.4) as each activity is introduced. These diagrams will assist the students as they work on the activities.

Activity A: Beam Bridge

Materials

- Plasticine
- several books
- 10 or more strips of Manila tag (or 3-ply Bristol board), cut into 15 cm x 45 cm pieces
- beam bridge diagram (4.5.1)

Instructions:

1. Make ten or more balls of Plasticine (about 2.5 cm across – a little smaller than a Ping-Pong ball)
2. Stack two equal piles of books (approximately 30 cm apart)
3. Lay one strip of Manila tag across the books to form a bridge.
4. Find out how many balls of Plasticine the bridge can hold.
5. Fold the Manila tag strips in different ways (e.g., crimp paper cross- or length-wise) to make different kinds of beams and test the beams for strength using the Plasticine balls.

Activity B: Arch Bridge

Materials

- scissors
- 2 strips of Manila tag (15 cm x 45 cm)
- ruler
- hardcover books of different sizes
- 10 balls of Plasticine
- arch bridge diagram (4.5.2)

Instructions:

1. Cut one strip of Manila tag so that it is 30 cm long.
2. On a flat surface, use this strip to make an arch 20 cm wide. Use books to support the arch on either side.

Note: The books should be as high as the arch.

3. Lay the other strip of Manila tag across the books and the arch to make the roadway of the bridge.
4. Put Plasticine balls on the roadway, one at a time, to test the strength of the arch.
5. Make other arches with narrower spans. Determine which bridge is the strongest.

Activity C: Suspension Bridge

Materials

- 2 identical chairs
- thick string
- 4 heavy books
- scissors
- large cardboard box
- packing tape
- suspension bridge diagram (4.5.3)

Instructions:

1. Place the chairs with their backs facing each other. Move the chairs apart so that there is a gap twice the length of the cardboard box. (The chairs are the bridge's towers.)

▶

2. Tie a piece of string from the top right-hand side of one chair to the top of the other. Do not pull the string tight. Leave it loose so that it forms a shallow curve. This is one of the cables of the bridge.
3. Repeat Step 2 with another piece of string attached to the top left-hand side of the chairs.
4. Press down on the two pieces of string and watch what happens to the chairs. (The chairs will move.)
5. See if you can find a way to use the heavy books and more string to stop the chairs from moving. (Use the diagram to help you.)
6. Cut off the long sides of the cardboard box. Take the sides and tape them together to make a roadway. Trim the roadway so that it is as wide as the chairs. (The rest of the box is unused.)
7. Attach loops of string that hang from one cable to another. The loops should be an equal distance apart.
8. Rest the roadway on the loops of string.

Activity D: Movable Bridge

Materials

- scissors
- cardboard box
- chair
- nail
- thick string
- packing tape
- 2 heavy books
- movable bridge diagram (4.5.4)

Instructions:

1. Cut off one side of the cardboard box. Trim the piece of cardboard so that its length is as wide as the chair seat.
2. Have your teacher use a nail to punch a hole near each corner at one end of the piece of cardboard.
3. Cut a piece of string. Push one end of the string through one hole. Tie a knot at the end of the string that is large enough so that the string won't pull back through the cardboard.
4. Cut another length of string and repeat step 3 with the other hole.
5. Tape the other end (the end without the holes) to the back of the chair seat.
6. Run the strings over the back of the chair and tie or tape each string to a book.
7. See if you can make the bridge lift and lower by moving the book.

Investigating Bridges

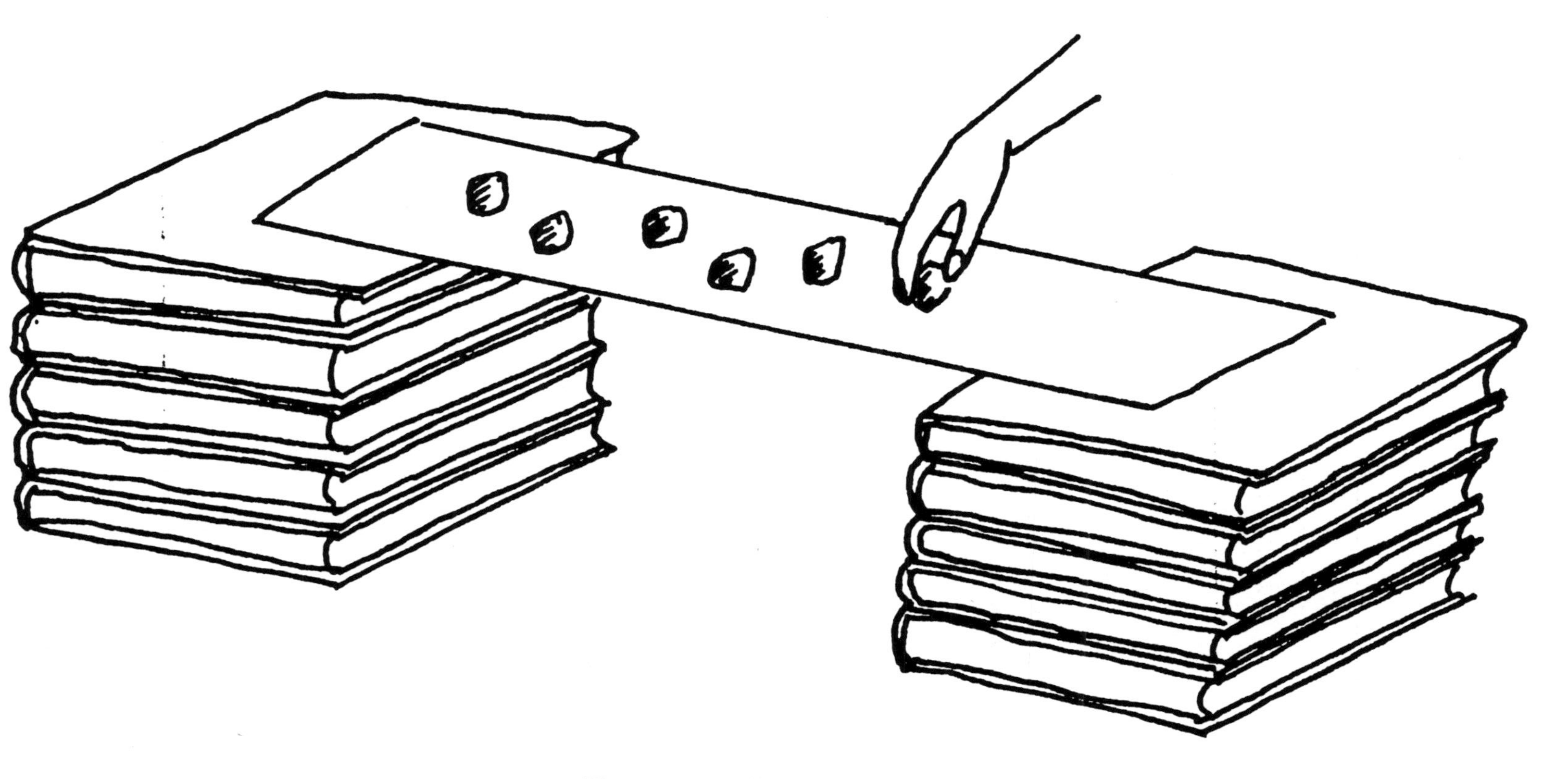

Beam Bridge

Investigating Bridges

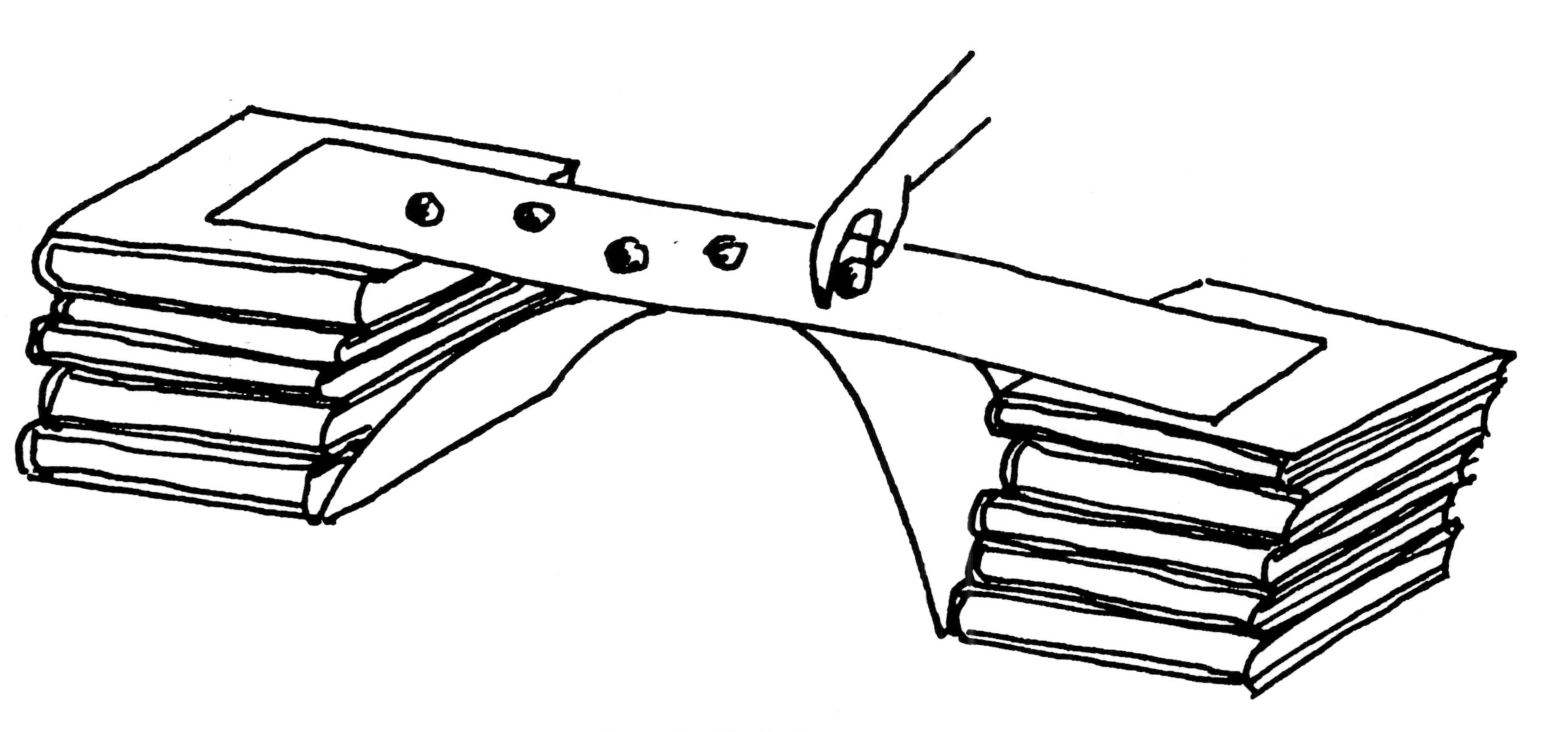

Arch Bridge

Investigating Bridges

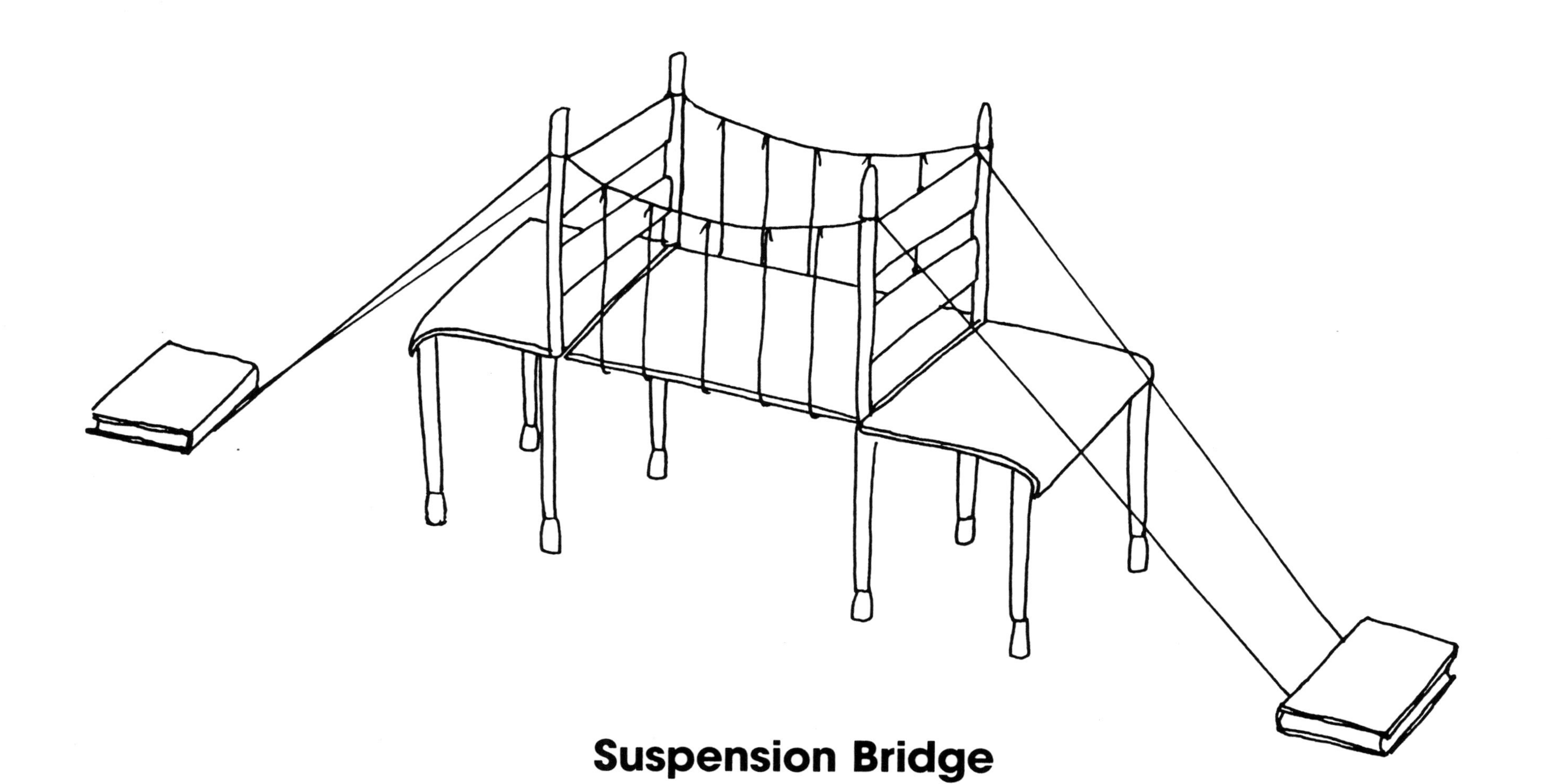

Suspension Bridge

Investigating Bridges

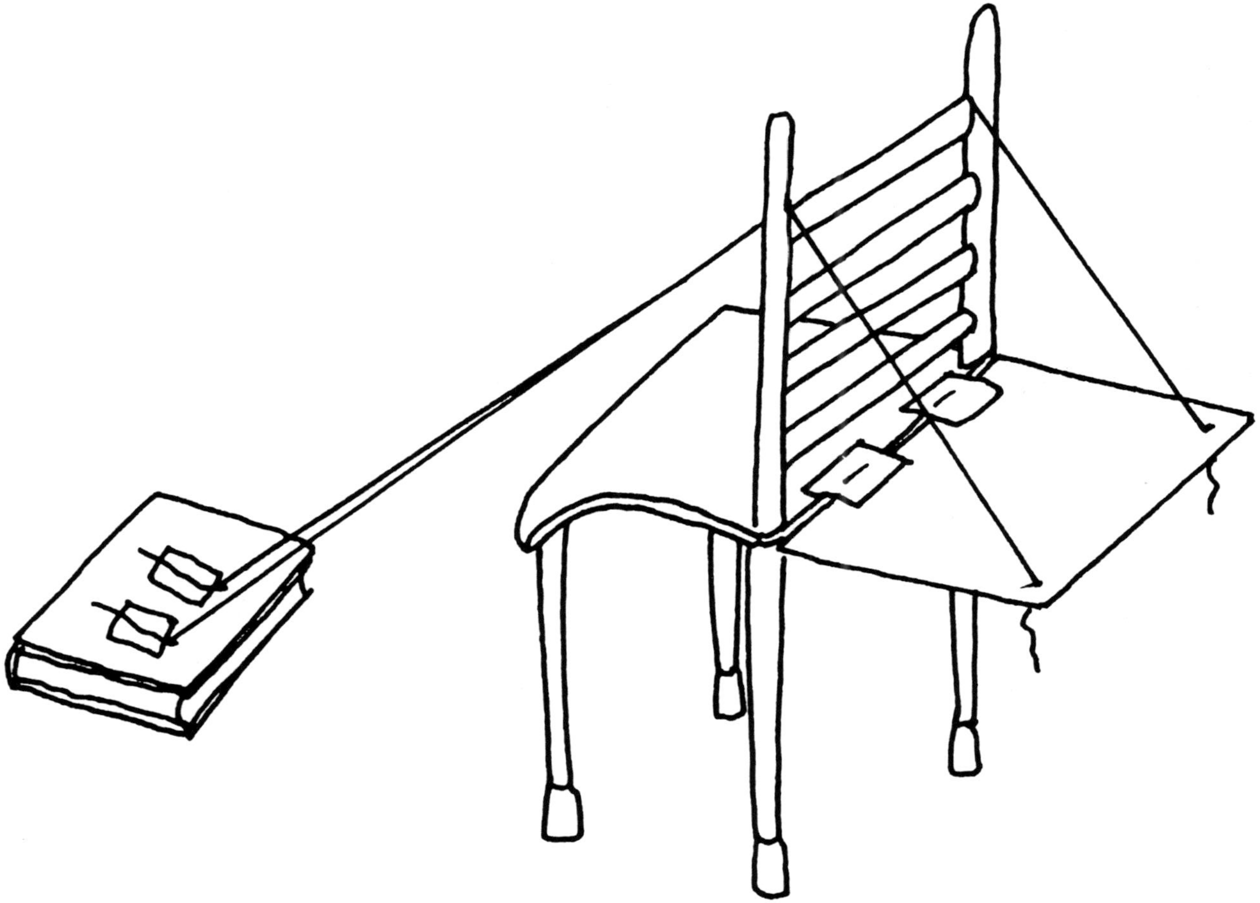

Movable Bridge

6 Designing and Constructing Bridges

Note: In this activity, students are given an opportunity to design and construct their own bridges using the ideas and experiences from previous lessons. Students should be encouraged to design unique bridge structures, as opposed to replicating a previously constructed bridge.

Materials

- variety of materials that students can use to construct their own bridges including:
  - string
  - Manila tag
  - straws
  - cardboard boxes
  - low-temperature glue guns
  - glue sticks
  - scissors
  - glue
  - paper fasteners
  - single-hole punch
  - packing tape
  - Plasticine

Activity

As a class, review the four different types of bridges studied (beam, arch, suspension, movable), discussing the design and ways that the bridges are made strong and stable. Divide the class into working groups and challenge the groups to design, construct, and present their own unique bridge.

Review the steps outlined below with the students.

1. In your group, decide which type of bridge you would like to construct. Record the name of the bridge on your activity sheet.
2. Design your bridge and draw a labelled diagram of it.
3. List the materials needed to construct your bridge.
4. List the steps you are going to follow to construct your bridge.
5. Gather the necessary materials.
6. Construct the bridge.
7. Test the strength of your bridge using Plasticine balls.
8. Decide on a name for your bridge.
9. Plan your presentation of your bridge to the class. Include the following in your presentation:
   - materials used
   - your initial sketch
   - description of how you built your bridge
   - discussion of the difficulties you ran into constructing your bridge
   - discussion of how you overcame your difficulties
   - demonstration of the strength of your bridge, using Plasticine balls

Provide the groups with plenty of time to design, construct, and test their bridge, as well as plan their presentations. During presentations, encourage the students to ask questions and provide positive feedback. You may also wish to make these presentations to other classes or set up a display for students, staff, and parents/guardians.

Activity Sheet

Note: The activity sheet is to be completed during the design and construction of the bridge.

Directions to students:

Use the activity sheet as a guide in designing and constructing your bridge (4.6.1).

Assessment Suggestion

During presentations, observe students' abilities to orally discuss their project and explain the designing and constructing process to their classmates. Use the anecdotal record sheet on page 15 to record results.

Date: ____________ **Name:** ____________________

____

____

____

Designing and Constructing a Bridge

1. **The type of bridge we would like to construct is:** __________

____

2. **Our Bridge Design:**

3. **Material Required:**

4. These are the steps we will follow to make our bridge:

____

____

____

____

____

____

____

____

____

____

____

____

5. Our bridge can support this many weights: ____________

6. The name of our bridge is: ______________________

7. Difficulties we had while designing and constructing our bridge were: ______________________

____

____

____

____

8. We overcame our difficulties by:

____

____

____

____

7 Structure and Mechanical Parts of Objects

Materials

- toy cars, airplanes, trucks, and so on

Note: Prior to this lesson, have the students bring in toy cars, airplanes, trucks, and so on from home. Display these toys on a table to build the students' curiosity.

- chart paper, felt pens

Activity

Divide the class into pairs or small groups of students. Distribute the toys so that each group has at least one toy. Give the students time to examine and manipulate their toy. Have them draw a labelled diagram of their toy on their activity sheet.

Once the students have had an opportunity to complete their drawing, have them share with the class some of the different parts of their toy. When each group has had a chance to discuss some of the different parts of the toys, discuss the difference between the structural parts and mechanical parts of the toys. Clarify the words *structural* and *mechanical* for the students.

structural: parts that give shape and stability to the toy

mechanical: parts that help make the toy work

Divide chart paper into two columns: Structural Parts and Mechanical Parts. As a class, brainstorm a list of parts of the toys that belong under each category. For example, the structural parts of a toy car include chassis (body), seats, and arm rests. The mechanical parts of a toy car include wheels, axles, motor, and steering wheel.

Students can now use this chart as a reference in classifying the labelled parts of their toy on their activity sheet.

Activity Sheet

Note: The activity sheet is to be completed during the activity.

Directions to students:

Draw a diagram of your toy and label its parts. On the chart, indicate whether the part is structural or mechanical (4.7.1).

Extensions

- Provide manipulatives such as Lego for the students to use to construct objects with structural and mechanical parts. Have students design their own transportation devices.
- Provide a collection of auto magazines for students to look through. Have them look in greater detail at the different components (both structural and mechanical) of a vehicle.
- Visit an automobile plant so that students can see firsthand the different components of a vehicle and how the components are put together to work as a system.

Assessment Suggestion

Have students complete a student self-assessment on page 22 to reflect on their own learning about structural and mechanical parts of objects.

Date: ____________ Name: ____________

Structural and Mechanical Parts

My object is a(n): ____________

Labelled Diagram:

Name of Part	Structural (X)	Mechanical (X)

References for Teachers

Boesveld, R., D. Needham, and A. Rhodes. *Simple Machines*. Burlington, ON: Halton Board of Education Curriculum Materials, 1989.

Bosak, Susan. *Science Is*. Kitchener, ON Encore Printing, 1986.

Carlisle, Norman, and Madelyn Carlisle. *I Want to Know About Bridges*. Chicago: Children's Press, 1972.

Carter, Polly. *The Bridge Book*. Toronto: Simon & Schuster, 1992.

Gibson, Gary. *Making Shapes*. Brookfield: Copper Beech Books, 1995.

Kaner, Etta. *Bridges*. Toronto: Kids Can Press, 1994.

Morgan, Sally, and Adrian Morgan. *Designs In Science – Structures*. London: Evans Brothers, 1993.

Oxlade, Chris. *Bridges*. Austin, TX: Steck-Vaughn, 1996.

Parkway Group. *Look!* London: Addison-Wesley, 1981.

Sturges, Philemon. *Bridges Are to Cross*. New York: GP Putnam's Sons, 1998.

Unit 5

Soils in the Environment

Books for Children

Baines, John, and Barbara James. *The Fragile Earth*. New York: Simon & Schuster, 1991.

Baylor, Byrd. *Everybody Needs a Rock*. New York: Scribner, 1974.

Bourgeois, Paulette. *The Amazing Dirt Book*. Toronto: Kids Can Press, 1990.

Hamilton, John. *ECO-Careers: A Guide to Jobs in the Environmental Field*. Edina, MN: Abdo & Daughters, 1993.

Hansen, Ann Larkin. *Crops on the Farm*. Minneapolis: Abdo & Daughters, 1996.

Johnson, Kipchak. *Worm's Eye View: Make Your Own Wildlife Park and Be Friends With Small Animals*. London: Cassell, 1990.

Munsch, Robert. *The Mud Puddle*. Toronto: Annick Press, 1982.

Patent, Dorothy Hinshaw. *Biodiversity*. New York: Clarion Books, 1996.

The Visual Dictionary of the Earth. Don Mills, ON: Stoddart, 1993.

Web Sites

- **http://www.uia.org/uiademo/pro/d0949.htm**

  Union of International Associations, from the Encyclopedia of World Problems and Human Potential: an educational site all about soil erosion, including natural and human-made causes, incidence, and background information. Extensive links are offered to both broader and narrower issues (such as environmental hazards from logging and loss of river silt).

- **http://net.indra.com/~topsoil/Compost_Menu.html**

  Welcome to Rot Web!: everything you need to know about composting, with teacher resources, links, and references.

- **http://129.128.49.169/Pedosphere/index.cfm**

  The Pedosphere and its Dynamics: A Canadian Perspective: an extensive site on soil, its components and properties – includes ecological functions of soil, texture, colour, and structure of soil, soil formation, mineralogy, soil water, soil air, and more.

- **http://www.nj.com/yucky/worm/**

  New Jersey Online "Worm World": an engaging source for teachers and students on worms as recyclers, all about earthworms, and further resources for teachers.

- **http://www.epsea.org/adobe.html**

  Adobe Home Construction for the El Paso Solar Energy Association: information on the construction of adobe houses, the materials used, and the structures' history. With links to straw bale homes, energy efficiency, and solar heating.

- **http://www.garden.org/edu/features.htm**

  American National Gardening Association's newsletter for teachers, titled "Growing Ideas – A Journal of Garden-Based Learning": scroll down to January 1997, "Sorting Out Soil: the Inside Dirt" for activities on soil and gardening in the classroom.

- **http://www.swifty.com/apase/charlotte**

  Association for the Promotion and Advancement of Science Education: click on "Forest Ecology" and then "Soils" to learn about this vital component of the whole forest ecosystem. An excellent site for teachers and students.

Introduction

Throughout this unit, students will conduct a number of hands-on investigations of soils in the environment. Through these investigations, students will discover that soils are made up of living things and different earth materials. Activities will look at different kinds of animals and plants that live in soil, as well as the importance of soil in providing a base for gardens, forests, fields, and farms.

Throughout the unit, students will investigate the similarities and differences between various soils. They will also look at the components of various soils, and describe the effects of moving water on these soils. In addition, students will identify the dependence of humans and other living things on soil, and recognize the importance of soil as a source of materials for making useful objects.

Some of the concepts in this unit will have been introduced in Unit 1, Growth and Changes in Plants. As a result, students may be familiar with some of the concepts and vocabulary. Their knowledge will only strengthen their confidence as they investigate new ideas throughout this unit.

This unit relies on outdoor activities for students to experience, firsthand, the importance of soil as a source of life and nourishment for many organisms; it should be taught during seasons when the ground is soft and the soil can be easily turned. Prior to teaching the unit, ensure that locations are identified near the school where investigations can take place.

To teach this unit, you will need many pictures, books, and films depicting different types of soil, growing environments for different plants, and living organisms that live in the soil. Involve students in this project. Sources for photographs and drawings are:

- old calendars
- magazines, especially *Ranger Rick, Owl, Chickadee, Highlights for Children*, and *National Geographic*
- departments of agriculture
- local garden nurseries
- gardening magazines
- seed catalogues
- forestry or environmental associations
- local farming or naturalist societies

Science Vocabulary

Throughout this unit, teachers should use, and encourage students to use, vocabulary such as: *clay, sand, pebbles, rocks, organic matter, sedimentation, sieving, loam, humus,* and *compost.*

Materials Required for the Unit

Classroom: magnifying glasses, rulers with mm markings, water, chart paper, felt markers, pens, pencils, masking tape, permenent markers, white glue, clipboards, scissors, felt pens

Books, Pictures, and Illustrations: pictures of different objects that use soil materials (for example, a brick building, a cobblestone walkway, shell/pebble mosaics, clay pots, clay jewellery), books on various types of fruits and vegetables

Household: glass jars with lids, paper towels, aluminum pie plate, funnels, plastic Ziploc bags, paper cups, plastic cups, spoons, small containers with lids (for example, yogurt containers), large bowl, empty 1-l milk carton (cut lengthwise), tweezers, large spoons

Other: garden trowel, shovels, newspaper, small container or jar filled with water, various soil samples (peat moss, potting soil, sand, pebbles, clay soil, and so on), Popsicle sticks,

clock or watch with a second hand, plastic jug or container, three different types of soil (good potting soil, vermiculite, peat moss, heavy clay, sand, gravel, and so on), radish or bean seeds or other relatively fast growing seeds, pots for planting seeds, small stones, small pebbles, grass clippings or small pieces of straw, Vaseline, wooden box with slats or compost bin, household or school garbage (for example, any remains from fruits and vegetables, grass clippings, leaves, paper), variety of small packets of seeds (carrots, radishes, lettuce, beans, and so on), index cards, watering can, spray bottle, labelled containers, sample bricks, dirt or clay, modelling clay, top soil, planting soil

1 Different Types of Soil

Science Background Information for Teachers

Soil is loose, broken down rock material in which plants can grow. Soil is a mixture of four components: mineral grains (sand, silt, and clay), organic matter or humus (the remains of once-living things), water, and air. The observable characteristics of some types of soil are:

Sandy Soil:

- falls apart easily
- feels gritty
- is mostly sand, mixed with a little silt, clay, and humus

Loam (garden soil):

- forms clods
- feels like velvet
- is about half sand, a third clay, and the rest is silt and humus

Clay:

- sticks together
- feels greasy
- is at least half clay; the rest is sand, silt, and humus

Humus:

- is soft and spongy
- falls apart quite easily
- is composed of decayed vegetation

Materials

- concept map (drawn on chart paper)
- plastic Ziploc bags
- garden trowels
- permanent markers
- chart paper, felt markers
- newspapers
- magnifying glasses
- tweezers
- containers of water

Activity

Note: Locate areas near your school where students can examine soil and collect samples (for example, in a garden, in the woods, along a road, beside a creek, in a vacant lot).

Display the concept map, and have students discuss and present what they know about soil. As they express their ideas, record these on the map.

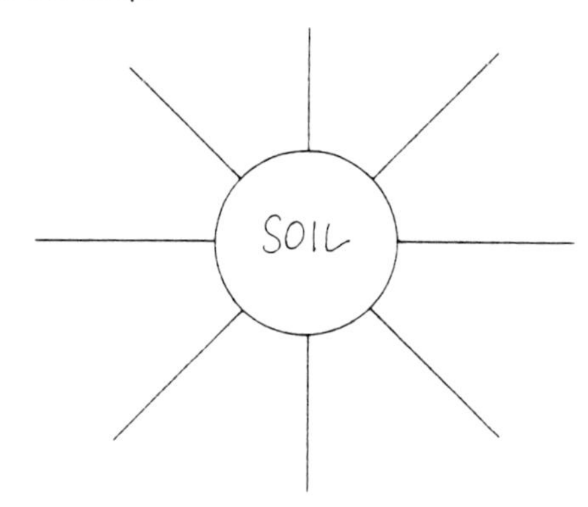

Tell students that, on a specified day, you are going to take them on a soil search. Encourage them to bring along old clothing, garden gloves, and rubber boots for the walk.

On the day of the soil search, divide the class into groups of three or four. Give each group four plastic Ziploc bags, a trowel, and a permanent marker. Take students on a walk to four different locations around the school neighbourhood. Have each group select a soil sample from the four locations and use the trowel to place the sample in the bag. Make sure students label the bag so that they know where the soil sample came from. At the end of the soil search, each group should have four different soil samples.

Back in the classroom, have students examine the bags of soil samples. Have them brainstorm a list of descriptive words on chart paper. For example, ask the students:

1

- What words could you use to describe the texture of the different soils?
- What words could you use to describe the smell of the different soils?
- What words could you use to describe the look of the different soils?

This list of descriptors will assist students in analyzing their soil samples.

Now have the students empty their soil samples into separate piles on newspaper. Make sure they keep the labelled bag near the pile of soil so they remember where the soil was taken from. Have students analyze their soil samples, using the magnifying glasses and tweezers. They may be able to recognize many different kinds of materials in the soil, such as bits of rock, sticks, leaves, seeds, sand, and clay. Encourage students to identify as many of these components as possible, and share their discoveries with others in their group. Have them record their observations on the activity sheet.

Have the students wet a small handful of soil by dipping their hands in a container of water and then holding the soil. Ask:

- How does the soil hold together?
- What does the soil feel like when you rub it between your fingers?

Continue having students record these observations on their activity sheets.

Note: Keep the soil samples from this activity for future activities.

Activity Sheet

Note: This activity sheet is to be completed during the activity. Make four copies of the activity sheet for each group of students.

Directions to students:

Record your observations of each soil sample (5.1.1).

Extensions

- Invite a horticulturist to speak to the students about different types of soils, the importance of soil for growing plants, and the different types of plants that grow in various soils.
- Visit a local nursery or botanical garden to observe the different types of soils used to grow different types of plants.
- Investigate how different communities or regions have different types of soil (e.g., prairie soil is different from coastal region soil). Contact schools or other organizations in different regions and request that they send you a bag of soil. Students can write the letters themselves. This is an excellent Language Arts and Social Studies connection.

Activity Centre

Encourage students to bring in soil samples from different locations (also include specific soil samples that you have collected, such as clay, humus, loam, and so on). Display the different soil samples at the activity centre. Using a magnifying glass, students can compare the various components of the different soil samples. Students can use additional copies of the activity sheet to record results.

Assessment Suggestion

Observe students as they examine and describe the soil types. Focus on their ability to describe the texture, appearance, and malleability of the soil samples. Use the anecdotal record sheet on page 15 to record results.

Date: ____________________ Name: ______________________________

Soil Search

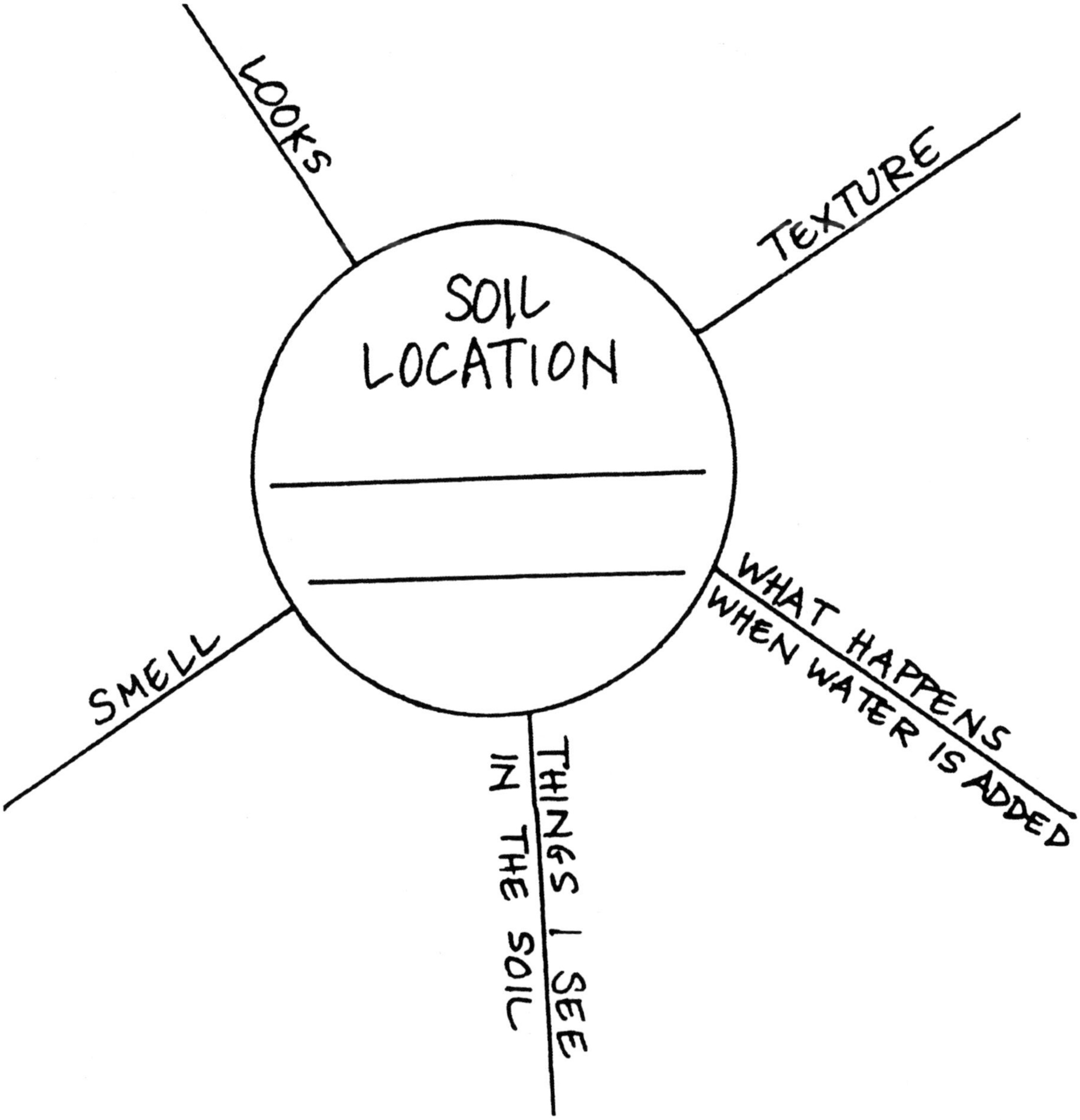

2 Components of Soil

Materials

- glass jars with lids
- samples of soil from different places and depths (one for each group) (Fill jars half full with soil. Make sure you place equal amounts of soil in the glass jars with lids. Label the jars with the location that the soil was taken from.)
- water
- spoons
- magnifying glasses
- record of observations
- newspapers
- paper towels
- aluminum pie plates
- ruler with mm markings
- funnels

Activity

Divide the class into small groups and have the groups cover their work surface with newspaper. Give each group a jar half filled with soil, magnifying glasses, some water, and a ruler. Have students dump out the soil onto a piece of newspaper and observe it with a magnifying glass. Ask the students:

- Can you see different bits and pieces in your soil?
- Can you sort them out?
- What do you think they are?

Next ask the students to put the soil back in the jars. Have them fill the jar with water, put the lid on it, and shake it twenty times. Tell the students to set the jar down on the desk and let it stand until the soil settles. Ask:

- Can you see layers in the jar?
- What is at the bottom of the jar?
- What other layers do you see?
- Is anything floating?

Have students walk around and observe the samples of soil from other groups. Ask:

- How do the different soils compare?
- Do they have the same layers?
- Do they have the same kinds of particles?

Note: The different layers in the jar are caused by different particles in the soil. Decaying materials – bits of wood, leaves, and roots – usually float on top of the water. The smallest particles dissolved in the water are clay. The fine grains are silt. The coarse grains are sand. The pieces of stone larger than the sand grains are pebbles.

Have the students draw a diagram of the jar on their activity sheet, as close to the actual size of the jar as possible (encourage them to measure the jar before drawing it). Now have them measure each layer in mm and draw the layers on the jar diagram. Have students label the layers using their own words to describe the layers.

Have the groups present their diagrams to the class. Compare the different samples. Ask:

- Are these soil samples the same?
- How are they different?

Activity Sheet

Note: This activity sheet is to be completed during the activity.

Directions to students:

Draw a diagram of your jar. Measure and draw the different soil layers inside it. Label the layers (5.2.1).

Extensions

- Remove the lids from the jars of soil samples and place the jars in a warm area. Leave them for several days so the water evaporates, then examine the layered soil samples again.

2

- Separate soil using sieving techniques. Provide the students with several sieves of varying sizes (include strainers, fine sieves, and even cheesecloth, mesh, nylon stockings, and screening). Have them separate soil samples by placing a sample in the sieve and shaking the sieve over a plate. Challenge the students to identify the components as the components are separated.

Date: ______________ Name: ______________________

Separating Soil

1. Draw a diagram of the jar.
2. Measure and draw the layers in the jar.
3. Label the layers.

What do you think each layer in the jar is made up of?

____

____

____

____

3 Absorption of Water

Materials

- 4 soil samples in labelled containers (such as peat moss, potting soil, sand, clay soil)
- paper cups
- plastic cups
- pencils
- Popsicle sticks
- paper towel
- water
- clock or watch with a second hand
- plastic jug or container

Activity

As a class, discuss how important it is for soil to be able to absorb water. Ask:

- Why is soil important?
- What is soil used for?
- What else do plants need to grow?
- How does a plant get its water from the soil?
- What do you think would happen if soil could not hold or absorb water?
- What is meant by absorption? (Use the example of a sponge soaking up water to illustrate the meaning of absorption, and compare the sponge to a brick in terms of its ability to absorb water.)
- Do you think all soil can absorb water equally well?

Divide the class into small groups. Give each group four soil samples. Have them examine each sample and record their observations on their activity sheet. Ask:

- Which soil sample do you think can absorb water very well?
- Which soil sample do you think cannot absorb water very well?

Explain to the students that they are going to test the ability of water to pass through the various samples of soil. Ask:

- Why would you want to know how much water passes through a soil sample?
- If all the water does not pass through the soil, where does it go?

Have the groups select one soil sample to test. Have the students use a pencil to poke ten holes in the bottom of a paper cup. Next have them place a folded piece of paper towel in the bottom of the paper cup to prevent granules from the soil sample from falling out.

Have the groups fill the paper cup half full with the soil sample. Remind students to label the paper cup with the name of the soil sample. Now have them set the plastic cup on a flat surface, and place two Popsicle sticks across the top of the plastic cup. The Popsicle sticks will form a bridge for the paper cup to rest on. This also allows water to flow from the paper cup into the plastic cup.

Note: It is important for students to measure time accurately during this investigation. You may wish to conduct the first few tests as a large group so students know exactly how to calculate the amount of time needed for the water to pass through a soil sample.

Have one student in each group act as a timer. This student's job is to watch the clock at all times in order to instruct the group when to start pouring the water into the paper cup and when to stop measuring the time. Beginning with the second hand at 12, have the student indicate to the group when the second hand reaches the 12 again, at which point another group member fills the paper cup to the top with water. The group then observes the cup and notes when the water begins to seep through the holes into the plastic cup. The timer then notes the time and the group

3

calculates the amount of time it took for the water to pass through the soil sample. This result is recorded on the activity sheet.

Have the groups repeat the investigation for each of the soil samples.

Following the experiment, ask the students:

- Which sample did the water seep through first?
- Which sample did the water seep through last?
- Did the same amount of water seep through each sample?
- How do you know?
- Did all of the water seep through?
- Where did the remainder of the water go?
- Which soil would be best for growing plants that need lots of water? Why?

Activity Sheet

Directions to students:

Examine each soil sample and record your observations. Record the amount of time it took before the water began to seep through into the plastic cup (5.3.1).

Extensions

- Have students devise a method for measuring the amount of water that seeps through each soil sample. Have them repeat the activity, using their measuring device to measure the water in the plastic cup.

- Have students construct a bar graph to compare how long it took before the water began to seep through into the plastic cup for each of the different samples. Label the graph "The Absorption of Water by Different Earth Materials."

- Read the book, *Mud Puddle,* by Robert Munsch. Discuss how soil is affected by rain water.

Assessment Suggestions

- As students conduct their investigations with water absorption in soil, observe the students' ability to work productively in a group. Use the cooperative skills teacher assessment sheet on page 21 to record results.

- Have students complete a cooperative skills self-assessment sheet on page 23 to reflect on their own ability to work together with classmates.

Date: ________________ **Name:** ____________________________

Investigation: Absorption of Water

Soil Sample #1 ______________________

Description ____

____

____

Length of time it took for water to seep through: ________

Soil Sample #2 ______________________

Description ____

____

____

Length of time it took for water to seep through: ________

Soil Sample #3 ______________________

Description ____

____

____

Length of time it took for water to seep through: ________

Soil Sample #4 ______________________

Description ____

____

____

Length of time it took for water to seep through: ________

4 How Different Soils Affect the Growth of Plants

Materials

- 3 different types of soil, such as potting soil, heavy clay, and sand
- radish or bean seeds, or other relatively fast-growing seeds
- pots for planting seeds
- small stones
- masking tape
- felt pens
- 3 charts for recording results (made prior to beginning this activity) (sample included) (5.4.1)

Activity

Display the seeds for the students to examine. Ask:

- What do these seeds look like?
- How could we find out what kind of plants these seeds will grow into?

Discuss the procedures for plant care. Ask:

- What do plants need in order to grow?
- Do you think the type of soil will make a difference to the way the seeds grow?

Have the students examine the types of soil. Ask:

- How are the types of soil different?
- What is each type of soil called?
- In which soil do you think the seeds would grow best?
- In which soil do you think the seeds would not grow well?

After putting a few stones in the bottom of each pot for drainage, have the students test their predictions by planting seeds in the various soil types. Label the pots according to the types of soil and seeds used.

Review plant location and discuss other factors that affect growth, such as access to sunlight and temperature. Also discuss the importance of creating a fair test when conducting an experiment, by ensuring that only the soil varies. The location and care should remain the same for all the plants. Have the students choose a good location in the classroom for the plants.

While the plants are growing, observe them every few days. Measure their growth and other changes and record this information on the charts.

For the duration of the experiment, have the students use their activity sheets as observation journals. Discuss the growth of the plants in terms of the soil in which they are planted. Have students make comparisons, identifying the best soil in which to grow the plants, and ordering the soil types from worst to best.

Activity Sheet

Note: Make three copies of the activity sheet for each student.

Directions to students:

Record the name of each soil type at the top of each chart, then print the observation dates and record information about the growth of the plants (5.4.1).

Extensions

- Have students graph the growth of their plants. Use the graph to draw conclusions about the types of soil that are best and worst for growing indoor plants.
- Take students for a walk around the schoolyard and local community. Have them observe the different types of plants that live in different types of soils. Discuss which types of plants require a specific type of soil in order to survive.

▶

4

- Visit a local nursery and discuss the composition of different types of soils used to grow different types of plants.

Assessment Suggestions

- Observe students as they measure the growth of their plants to determine if they can measure accurately using a cm ruler. Use the anecdotal record sheet on page 15 to record results.
- Following the experiment, have students complete a student self-assessment on page 22 to reflect on their learning about plants and soil.

Date: ____________ Name: ____________________

Plant Journal

Type of Soil: ____________________

Date	Observations

5 The Characteristics of Soils and Plants

Science Background Information for Teachers

The root systems of plants are classified as either tap roots or fibrous roots. Tap roots are thick, singular roots such as those found on carrots, dandelions, and radishes. Fibrous roots are somewhat stringy and have a multiple system of roots such as those found on marigolds, beans, and corn plants.

Materials

- variety of small packets of seeds, including those with fibrous roots and tap roots (include carrots, radishes, lettuce, beans, and so on)
- books on various types of fruits and vegetables
- planting pots
- potting soil

Activity: Part One

Distribute the packets of seeds to the students. Have them read the packets carefully. Ask the students:

- How do you know where these seeds should be grown?
- What type of soil is suggested for these seeds?
- Why is it important to know the type of soil needed before growing these plants?
- What other important growing factors are outlined on your seed packet?

Have students compare the growing instructions for the different seed packets. Encourage them to take note of the similarities and differences.

Activity Sheet A

Note: Students can refer to the seed packages and plant books to complete the activity sheet.

Directions to students:

Select two plants that grow in your local environment. Write the name of the plant at the top of the sheet and draw a picture of the plant. At the bottom of the page, list the type of soil and the characteristics of the soil that are best for the plant. You may also wish to include any additional factors that may affect the growth of the plant (5.5.1).

Extensions

- Have students work in small groups. Distribute books and resources gathered on different types of plants. Have students use the resources to find where different types of plants are grown. Place a map of the world in the centre of a bulletin board. Have students write the name of one of the plants they have researched on an index card. Display the cards around the outside of the map. Have the students locate where the crop is grown on the world map. Attach a piece of string from the index card to the map. Use a pushpin to hold the string in place at the designated location. Label the bulletin board Plants From Around the World.
- Have students visit local farms to see what types of crops are grown, and in which types of soil the crops are grown.
- Invite a local farmer or produce manager to the class to talk about different crops in the local area.

Activity: Part Two

Note: Root systems of plants are introduced in Unit 1, Growth and Changes in Plants.

Note: This activity will take place over a long period of time, since students are growing plants from seed to adult.

Have students plant two types of seeds from the packages in the recommended type of soil. Ensure that they grow one plant with a fibrous root and one plant with a tap root. Once plants have grown to an adequate size, have students examine the plants. Then have the students remove the plants from the pots and shake off the soil to display the root systems. Have students examine the roots and draw a diagram of each plant and its root system on the activity sheet. Ask:

- How are the roots the same?
- How are the roots different?
- How do the roots grow differently in soil? (Tap roots grow straight down; fibrous roots spread out in several directions.)
- What special role do roots play in a plant's survival?

Re-introduce the terms *fibrous roots* and *tap roots* and encourage the students to use these terms as they discuss the plants and complete the activity sheet.

Activity Sheet B

Directions to students:

Draw a diagram of each plant and its root system. Describe the difference between a tap root and a fibrous root (5.5.2).

Extension

As a class, dig up several weeds in the school playground or surrounding area. Note how easy or difficult it is to pull out or dig up the weeds, and relate this to the type of root system found on the plant. Examine the weeds in terms of their root systems, where they were found, and the type of soil they were growing in.

Date: ________________ Name: ____________________________

Plants and Soil

Plant #1 ____________________________

Type of soil required to grow plant: ____________________

Characteristics of the soil: ______________________________

____

Other things that will affect the growth of the plant: ____

____

____

Plant #2 ____________________________

Type of soil required to grow plant: ____________________

Characteristics of the soil: ______________________________

____

Other things that will affect the growth of the plant: ____

____

____

Date: ______________ Name: ______________________

How Plant Roots Grow Through Soil

Plant With Tap Root	Plant With Fibrous Root

How are a tap root and a fibrous root the same? ______

____

____

How are a tap root and a fibrous root different? ______

____

____

____

How does a tap root grow through soil? ______________

____

____

How does a fibrous root grow through soil? __________

____

____

6 Investigating Living Things Found in Soil

Science Background Information for Teachers

Many living things survive in soil. Some living things, like earthworms, dig tunnels in the soil. This action allows air and water into the soil and that helps plants grow. Worms eat dried-up leaves and dirt. When they consume dirt, it passes through their body leaving rich, fertile nutrients behind.

Other living things that live in soil include various insects such as cutworms and slugs.

Materials

- newspaper
- shovels
- large plastic Ziploc bags
- containers of water (large milk jugs)

Activity

Note: Select a site for a field trip where you will be able to investigate soil samples and living things that survive in the soil.

Explain to students that they are going to investigate living things that are found in soil. Ask:

- What types of living things do you think you will see?

Divide the class into small groups and provide them with shovels and large plastic bags. At the site, have each group dig a deep hole in the ground. As they dig, have them look for animals: worms, insects, spiders, and so on. Use the bags to collect some of the soil and animals (if possible).

When students have dug down about 30 cm, draw their attention to the sides of the hole. Ask:

- What do you notice about the soil as you get deeper into the ground?
- Does the colour of the soil change?
- Does the texture of the soil change?
- Is the soil wetter or drier the farther down you go?
- Is it harder or easier to dig?

Next have the students pour some water into the hole and observe. Ask:

- Does any of the soil float? Which parts?
- Where do you think the water goes?

If possible, repeat the activity at another site so that students can draw comparisons.

When you return to the classroom, ask the students to describe all the different living things they saw in their pile of dirt. Examine the soil samples to see if there are any living things still in it. Have them identify the living things (if they know the names of them), or describe them. Have them use the activity sheet to describe and draw diagrams of the living things they have observed.

Activity Sheet

Directions to students:

Describe and draw diagrams of each living thing you found in the soil (5.6.1).

Extensions

- Use reference books to identify the names of the living things found in the soil samples. Have students select one living thing and research it, including a diagram and written information. Encourage them to determine how the living things affect the soil or the plants that live in the soil. Collect students' research papers and bind them together in a class book called *Creepy*

Crawlies and Other Things Found in Soil. Display the book in the classroom or the school library.

- Have students investigate earthworms in greater detail (for example, they can use a magnifying glass to look at the segments on a worm's body, observe how worms move through soil, and design experiments to determine what worms eat).
- Create a home for some living things that are found under the ground (earthworms, snails, ants, centipedes, and so on). Have students use an aquarium to recreate their natural environment. Fill the aquarium with soil, plants, and rocks.
- Have students design and draw a mural of life underground. Encourage them to incorporate all of the different living things they found during their soil investigation.

Date: ______________ Name: ______________________

Living Things in the Soil

Diagram	Description

7 The Effect of Moving Water on Soil

Materials

- sand
- white glue
- spoons
- small containers with lids (e.g., yogurt containers)
- small pebbles
- water

Activity

Begin the activity by asking the students if they have ever built a sandcastle on a beach. Ask:

- Would your sandcastle still be there now? Why not?
- What happens to the sand as the waves roll in?

Explain to the students that water from oceans, lakes, and streams affects many things in different ways. Ask:

- What do you think happens when water runs over rocks for a long period of time?
- How long do you think it takes for a rock to be worn away?

Explain to the students that they are going to simulate how rocks are broken down by water. Divide the class into small groups. Have each group mix about eight spoonfuls of sand with a large amount of white glue. Once the sand and glue are mixed, have the students shape the mixture into cubes and allow each cube to dry for two days.

After two days, have the students observe the cubes and record a description of the cube on their activity sheet. Have each group put its cube inside a small container with a lid, along with some small pebbles and water. Have the students take turns shaking the container. Each container should be shaken for approximately five minutes.

After five minutes, have the students remove the lid and observe the changes that have taken place. Have them record their observations of the cube and of the water in the container on their activity sheet. Ask:

- Why is the cube smaller than when you put it in the container?
- What does the water in the container look like?
- How are these results similar to what you would see in nature (e.g., at a beach, in a river or stream)?
- What happens to rocks and pebbles in a river over time?

Activity Sheet

Note: Observations should be recorded during the experiment.

Directions to students:

Record your observations from the experiment. Draw a diagram of the cube and describe the cube before it is shaken. Draw a diagram of the cube and describe the cube and the container of water after the cube was shaken in the plastic container with the water and pebbles. Answer the questions at the bottom of the sheet (5.7.1).

Extensions

- If your school is located near a beach, stream, or river, take the students to observe, firsthand, the effects of moving water on sand, rocks, and pebbles.

  Note: Discuss water safety before taking children near water.

  Discuss others ways in which nature changes rocks (e.g., erosion, glaciers).

- Read *Everybody Needs a Rock*, by Byrd Baylor. Have students select their own

▶

favourite rock or pebble. Have them name their "pet" rock and tell the many adventures of the rock (e.g., how it got its shape, where it lived). Have the students write a journal entry for their pet rock, in the first person.

- Have students put together a small booklet describing the various steps rocks go through as rocks break down from larger to smaller pieces.

Activity Centre

Have students bring in rocks and pebbles of varying sizes, shapes, and colours. Spread them on a large table and classify them. Have the students select the classification method.

Date: ______________ Name: ______________________

Effects of Moving Water

Sand/Glue Cube	______ ______ ______ ______ ______ ______
Cube in Container After Shaking	______ ______ ______ ______ ______ ______

Why did the cube change when it was shaken in a container of water and pebbles? ______________

____

How is what happens in this experiment similar to what happens to rocks and shells in rivers and oceans? ______________

____

8 The Effects of Rainfall on Surfaces

Materials

- asphalt
- lawn
- garden bed or bag of topsoil
- watering can
- spray bottle
- water
- clipboards
- pencils

Activity

Explain to the students that they are going to investigate the effects of water on different surfaces: asphalt, grass, and soil. Ask:

- How do you think rainfall would affect asphalt (grass/soil)?

Gather the required materials (watering can, spray bottle, and water), and have the students put their activity sheet on a clipboard to bring outside with them.

Activity A

Take students to a blacktop (asphalt) section of the playground. Have them make a circle so that they can all see. Ask:

- What do you think will happen when water from this bottle is sprayed on the asphalt?

Have them record their prediction on their activity sheet.

Spray the asphalt using the spray bottle. Ask the students:

- What happened when I sprayed water on the asphalt?

Make sure students have an opportunity to take a close look at the asphalt. Have them record their observations on their activity sheet.

Now ask the students:

- What do you think will happen when I pour water out of the watering can on the asphalt?

Have them record their prediction on their activity sheet.

Pour water on the asphalt. Have students take a close look at the asphalt and record their observations on the activity sheet.

Activity B

Have students make a circle around a patch of grass in the schoolyard. Have students take a close look at the grass before starting the investigation. Ask:

- What do you think will happen when I spray the grass with the water from the spray bottle?

Have the students record their predictions on their activity sheet.

Spray water on the grass. Once the students have looked at the grass closely, have them record their observations on the activity sheet.

Next ask the students:

- What do you think will happen when I pour water out of the watering can on the grass?

Have the students record their prediction on their activity sheet.

Pour water on the grass. Have students take a close look at the grass, then record their findings on the activity sheet.

Activity C

Take students to a garden bed on school property. If you do not have a garden bed with good topsoil, empty a bag of topsoil onto a garden bed or some grass. Ask the students:

- What do you think will happen when I spray water from the spray bottle on the soil?

Have the students record their prediction on their activity sheet.

Spray water on the soil. Have the students closely examine the soil and record their findings on their activity sheet.

Next ask the students:

- What do you think will happen when I pour water from the watering can on the soil?

Have the students record their predictions on their activity sheet.

Pour water from the watering can on the soil. Have the students take a close look at the soil and record their findings.

Once the students have returned to the classroom ask:

- Using the results from your investigations, what do you think the effects of a light rain shower would be on asphalt (on grass, on soil)?
- What do you think the results of a heavy rain downpour would be on asphalt (on grass, on soil)?
- How would other living things in the soil be affected by a light rain shower?
- How would other living things in the soil be affected by a heavy rain downpour?

Activity Sheet

Note: This activity sheet is to be completed during the activity.

Directions to students:

Record observations before and after each investigation (5.8.1).

Extensions

- Conduct investigations on other natural events (e.g., wind and ice storms) and their effects on soil.
- Have students investigate how different landforms (e.g., waterfalls, canyons, mountains) have been affected by natural events.
- Take students on a walk around the neighbourhood after a rainstorm. Have them observe the effects of the rainstorm on different surface areas (e.g., asphalt, grass, creek, garden bed).

Date: ____________________ Name: ________________________________

The Effects of Water on Different Surfaces

Surface	Water from Spray Bottle		Water from Watering Can	
	Prediction	Result	Prediction	Result
Asphalt				
Grass				
Topsoil				

9 Recycling Organic Materials

Materials

- shovel
- soil
- poster paper
- pencil crayons
- wooden box with slats (compost bin)
- household or school garbage (e.g., any remains from fruits and vegetables, grass clippings, leaves, paper)
- chart paper headed with Directions for Making a Compost Heap:

Directions for making a compost heap:

1. Shovel a layer of soil into an old wooden box with slats.
2. Add a layer of garbage (e.g., grass clippings, leaves, vegetable peelings, fruit cores, and water).
3. Cover the garbage with another layer of soil and mix.
4. Leave the composter outside where the rain will keep it moist.
5. Keep adding soil and garbage and mix around.
6. In about two months, use the humus from the composter for garden beds on the school property.

Activity

Begin by reviewing the various types of soil that students have studied and investigated throughout the unit. Ask:

- What kinds of soil can you name?
- Which types of soil are good for a vegetable garden?
- What could you add to the soil to make it more fertile to grow plants in?
- What is humus?
- What is humus made of?
- Can you think of a way that you could make your own humus?
- Have you ever seen a compost?

Explain to the students that people often build their own compost heaps so that they can produce fertile topsoil or humus for their gardens.

Display the directions for making a compost heap, and have the students construct the compost heap in the bin. Encourage them to collect organic garbage from home to put in the bin.

When the humus is ready to use, encourage students to mix it with soil for plants in the classroom (or they could take some home for their houseplants and gardens).

Now challenge the students to share their knowledge about composting by creating a poster or brochure to encourage members of the local community to compost. As a class, decide on five criteria for the poster or brochure; for example, it must include:

- an appropriate title
- a definition of what composting is
- an explanation as to why composting is important
- information on how to make a compost
- it must be attractive and easy to read

Display the posters and brochures in the school and around the local community. Also encourage students to share this information with their families and begin a compost heap at home.

Extensions

- Challenge students to use the design process to design and construct their own simple composters. Have students test and compare their composters to determine their effectiveness.

- If you do not have a school compost program in effect, you may wish to initiate this activity by having students organize one. Place plastic buckets in classrooms (or lunchrooms) with the word "Compost" clearly marked on them. When students are knowledgeable in the process and purpose of composting, have them visit classrooms to teach other students about the importance of composting, and to introduce the composting program. Students can make posters to advertise the new program.
- Have the students write letters to local government officials responsible for the environment, to request information about composting programs. Include copies of their brochures and posters with the letters.
- Set up a worm composter to investigate how these living things affect decomposition.
- Compare different types of soils to identify decomposers. Fill two containers with soil – one sterilized potting soil and the other soil from a local garden. Moisten both with water and place lettuce leaves on top of each type of soil. The lettuce leaves in the container of garden soil will decompose faster because of the living things within the soil. The sterilized potting soil has no living things in it to speed up the decomposition process.
- Visit a local mushroom farm to learn about the importance of compost in the production of mushrooms.

Assessment Suggestion

Use the criteria for the brochures and posters developed by the class on page 19. Record results for each student's poster or brochure on the rubric.

10 Using Soil Materials to Make Useful Objects

Materials

- printed chart with instructions for making bricks
- sample bricks
- large bowl
- dirt or clay
- water
- grass clippings or small pieces of straw
- scissors
- empty 1-litre milk cartons cut lengthwise (to use as brick moulds)
- scissors
- Vaseline
- large spoons
- objects or pictures of objects that use soil materials (e.g., a brick building, a cobblestone walkway, clay pots, soapstone carvings, sod houses)
- chart paper, felt pens
- modelling clay

Activity: Part One

Display the items or pictures of different objects that are made of soil materials. Ask the students:

- What do these items have in common?
- What are these things made from?

Explain to the students that soil materials are used to make many useful objects. Have the students brainstorm a list of things made from soil material. Record their answers on chart paper.

Focus on the sample brick. Ask:

- What are bricks used for?
- How do you think bricks are made?

Explain to the students that they are going to have an opportunity to make their own bricks. Review the directions for making bricks:

1. Put a couple of handfuls of dirt or clay into the large bowl.
2. Add water until the dirt/clay looks like pancake batter.
3. Add grass and straw to the mixture to bind it together.
4. Grease the mould (milk carton cut lengthwise) with Vaseline.
5. Spoon the wet mixture into the mould until it is full. Make sure it is even by thumping the mould on a hard surface a few times.
6. Set the mould in the Sun to dry.

Note: It will take a few hot, dry days to bake the bricks. They can be placed in the classroom on a window ledge in the sunlight, or above the heat register.

When the bricks are dry, have students compare their bricks to bricks from houses and other buildings.

As a class, build a structure using all of the bricks made by students.

Activity: Part Two

Challenge the students to design and construct a useful object made from modelling clay. Have them use the activity sheet to design their object, then make it and present it to the class.

Activity Sheet

Directions to students:

Design a useful object made from clay. Draw a diagram of your design, list the materials you will need, and describe how you made your object (5.10.1).

Extensions

- Visit a local potter to see how clay and pottery items are made.

▶

10

- Have students cut out pictures from magazines and newspapers of items that are made of soil materials.
- Have students conduct research projects on the various ways different cultures use earth materials (for example, clay pots, sod houses, adobe bricks, and soapstone carvings).

Assessment Suggestion

Have the students present their project idea to the class. You may wish to develop a class rubric as to how the presentation will be evaluated (for example, voice quality, eye contact, and clarity of instructions). List these criteria on the rubric on page 19 and record results as students present their projects.

Date: ____________________ Name: ________________________________

Designing a Clay Object

I will design a ______________________________.

I will need these materials:

______________ ______________ ______________

______________ ______________ ______________

______________ ______________ ______________

Making my clay object:

____

____

____

____

References for Teachers

Albert, Toni. *Rocks & Minerals*. Greensboro: Carson-Dellosa Publishing, 1994.

Bourgeois, Paulette. *The Amazing Dirt Book*. Toronto: Kids Can Press, 1990.

Butzow, Carol, and John Butzow. *Science Through Children's Literature*. Englewood: Teacher Ideas Press, 1989.

Fredericks, Anthony, and Dean Cheesebrough. *Science for All Children*. New York: HarperCollins, 1993.

Hoover, Evalyn. *The Budding Botanist - Investigations With Plants*. Fresno: AIMS Educational Foundation, 1993.

Pinet, Michele, Alain Korkos, and Fay Greenbaum. *Be Your Own Rock and Mineral Expert*. New York: Sterling, 1997.

Wyler, Rose. *Science Fun With Mud and Dirt*. New York: Simon & Schuster, 1986.